This book is about doing research, not about the results obtained. Those engaged on their first research may have had plenty of preparation about the techniques and results of prior research related to their proposed study, but may have limited knowledge of the actual strategies employed, or pitfalls encountered, by others who have conducted successful field and survey studies.

In this book, a number of researchers with experience of working on problems including environmental stresses, population genetics, parasitic vectors and vital records, describe obstacles encountered and successful strategies used in their own studies and by those of others. One learns to do research by trial and error, but accounts by experienced investigators can supplement what one learns from mentors and fellow students.

It is hoped that this book will prove helpful to biological anthropologists and will have applicability in related research pursuits.

T0321603

Cambridge Studies in Biological Anthropology 13

Research strategies in human biology

Cambridge Studies in Biological Anthropology

Series Editors

G. W. Lasker
Department of Anatomy, Wayne State University,
Detroit, Michigan, USA

C. G. N. Mascie-Taylor
Department of Biological Anthropology,
University of Cambridge

D. F. Roberts
Department of Human Genetics,
University of Newcastle-upon-Tyne

R. A. Foley
Department of Biological Anthropology,
University of Cambridge

Also in the series

Research strategies in human biology: field and survey studies

EDITED BY

G. W. LASKER
Department of Anatomy and Cell Biology,
Wayne State University, Detroit, Michigan, USA

AND

C. G. N. MASCIE-TAYLOR
Department of Biological Anthropology,
University of Cambridge, Cambridge, UK

CAMBRIDGE
UNIVERSITY PRESS

CAMBRIDGE UNIVERSITY PRESS
Cambridge, New York, Melbourne, Madrid, Cape Town, Singapore, São Paulo

Cambridge University Press
The Edinburgh Building, Cambridge CB2 2RU, UK

Published in the United States of America by Cambridge University Press, New York

www.cambridge.org
Information on this title: www.cambridge.org/9780521431880

First published 1993
This digitally printed first paperback version 2005

A catalogue record for this publication is available from the British Library

ISBN-13 978-0-521-43188-0 hardback
ISBN-10 0-521-43188-3 hardback

ISBN-13 978-0-521-01909-5 paperback
ISBN-10 0-521-01909-5 paperback

Contents

Boxes

Contributors

B. Bogin
Department of Behavioral Sciences, University of Michigan at
Dearborn, Dearborn, MI 48128, USA

G. W. Lasker
Department of Anatomy and Cell Biology, Wayne State University,
504 East Canfield Avenue, Detroit, MI 48201, USA

P. W. Leslie
Department of Anthropology, SUNY at Binghampton, Binghampton,
NY 13901, USA

M. A. Little
Department of Anthropology, SUNY at Binghampton, Binghampton,
NY 13901, USA

C. G. N. Mascie-Taylor
Department of Biological Anthropology, University of Cambridge,
Downing Street, Cambridge CB2 3DZ

J. H. Mielke
Department of Anthropology, University of Kansas, Lawrence,
KS 66045, USA

D. F. Roberts
Department of Human Genetics, University of Newcastle upon Tyne,
19 Claremont Place, Newcastle upon Tyne NE2 4AA

S. S. Strickland
London School of Hygiene and Tropical Medicine, Keppel Street,
London WC1E 7HT

A. C. Swedlund
Department of Anthropology, University of Massachusetts, Amherst,
MA 01003, USA

S. J. Ulijaszek
Department of Biological Anthropology, University of Cambridge,
Downing Street, Cambridge CB2 3DZ

Preface

This book is about doing research, not about the results obtained. Those engaged in their first research and, indeed, some with more experience, may have had plenty of preparation about the techniques and results of prior research related to their proposed study, but more limited knowledge of the actual thoughts and activities of others who have conducted successful field and survey studies of similar kinds. One can also profit by learning of obstacles encountered in the conduct of such research.

In most sciences a sort of apprentice relationship to a teacher helps span this gap. In biological anthropology, however, the problem of becoming an independent investigator may be exacerbated by the loneliness of field work far from one's university and the support of mentors and other fellow researchers there. Thus preparation for research of this kind may benefit from accounts about doing research by a number of authors with various kinds of research experiences. A few representative examples have been selected to illustrate the range of studies. To increase further the number of points covered, brief boxes with further examples from other studies are interpreted with the text.

This is not a methods book with details on how to do measurements and other procedures. Such methods, whether long-standing well-standardized anthropometry or new techniques devised for special purposes, are usually presented in the 'methods' sections of research reports, but the problem of research strategies, the subject of this work, is often omitted. It is hoped that a book on this aspect may prove helpful to biological anthropologists and have some applicability in related research pursuits by others.

We thank our co-editors of Cambridge Studies in Biological Anthropology and the staff for the biological sciences at Cambridge University Press for help and encouragement, and anonymous critics for concrete suggestions.

G. W. Lasker
C. G. N. Mascie-Taylor

1   *Planning a research project*

GABRIEL W. LASKER

Science advances by the development of new observations and concepts. Planning experiments and observations that may lead to changes in theory is not easy because these advances often depend on critical observations and interpretations that involve imponderables. Thus relatively small studies are sometimes the ones that trigger large advances in understanding, whereas in other instances they merely fill small gaps in accepted knowledge. When conceptual paradigms change, new tests or ways of interpreting old ones may ensue. Late in a very successful career in reproductive endocrinology, Parkes (1985) concluded that research can be planned, but discovery not; it is the thinking of the investigator rather than the technique of investigation that matter the most.

At risk of oversimplification, one can group many notions about scientific advance into one or the other of two contrasting ideas: 1. methods (finer and more exact measurements) may lead to new theories; 2. moving from one hypothesis and test to a new hypothesis is the key to scientific advance. There seems always to be an element of unplanned serendipity associated with major advances, but this usually follows carefully planned tests of prior ideas and attempts to confirm or contradict them. Thus the testing of prior ideas forms the platform of knowledge on which new ideas are constructed. Various theories of science all lead to the conclusion that planning one's studies is desirable.

Biological anthropology is a subject of wide scope. The research through field studies and surveys to be discussed in this book deals primarily with one aspect of it: human biological variation. To understand the causes and consequences of the variation, it must be dealt with both within and between human groups. Studies must not only consider relationship with other biological variables, but also the relationship with climatic, geographic and other physical conditions, and with social and cultural influences. This identifies such studies as anthropology, although those engaged in the studies may not be trained as anthropologists and the skills they contribute from other disciplines are often essential. Whatever their background, those embarking on such studies who have not had formal training in anthropology should include in their prep-

1

aration some reading of general works on the subject such as Harrison *et al.* (1988).

Choice of topic

In reports of research, attention is usually focussed on the hypothesis to be studied but not on the rationale for the choice of the general subject matter. That is because the subject matter may be determined by someone other than the researcher such as a professor or employer. In addition to the competencies and interests of the researchers, availability of funds and other resources may influence the decision. Nevertheless, the choice of the general subject and specific topic is the first important task in any research project. The limitations of available material, accessibility of populations, facilities and the prior training of the researchers, are primary considerations, but the relevance of the work for other people, including the general public, also influence the choice.

There is a difference between a broad subject of interest and a narrower topic for specific research. The broad area is explored by appropriate reading and study. The narrower topic should relate to this knowledge in some meaningful way if research is to be more than a disconnected set of items for the questions of a quiz show host. The topic is best set out in the form of a hypothesis or question to be answered by results of the proposed procedures of the study. Thus posed in operational terms and scaled to the limitations of the circumstances, there should be an excellent prospect of tangible results that can be published as a research paper (see Chapter 8).

It is possible to embark on field work or surveys with more than one hypothesis in mind during the data collection phase of the study. Nevertheless, the strategy must be specified in the plan. What observations are to be made, and how will the results be analyzed? Hypotheses should not be intermingled in a way that will make it difficult to know where to start. Failure to plan may make it difficult to interpret results in the light of results of prior work. If subsidiary issues are of real concern, one must consider whether they are an integral part of this study or should be reserved for the next.

That the general subject is biological anthropology in its wide sense is taken for granted by the frame of reference of this book. That still leaves a vast variety of subjects. There is no generally accepted catalog of subspecialties of biological anthropology nor has the field clearly defined limits. However, the focus of this work is on field studies and surveys, hence on living peoples or the recently dead rather than on old bones and

Table 1.1. *Conditions related to human biological variation*

1. Aspect
form, anatomy   function, physiology   body composition   biochemistry
growth   genetics   response to stress   diseases

2. Anatomical system
bones   teeth   joints   muscles   vessels   nerves   organs

3. Environment
climate   altitude   occupation   diet   housing and education
social relations   pollutants

4. Geographic focus
continents   nationalities   ethnic groups   isolates

5. Sex
female   male

6. Age group
embryo and fetus   neonatal and infancy   preschool   preadolescent
adolescent   reproductive   post reproductive

fossils. Furthermore, the concern here is with population biology rather than with individuals as such.

Research in biological anthropology may be experimental, descriptive, or comparative. Some features, such as statistical analysis, may apply to each of these, but laboratory experiments are generally in areas of overlap with other sciences, and description, which was characteristic of earlier phases of the subject, is now largely overlaid by the comparative approach. Comparisons of the very characteristics that had been described serve as a form of experimental approach where the experiments have been performed by nature. The anthropologist may select for study a comparative situation that it would be impractical or unethical to create in a human experiment. The task of planning may be viewed as largely one of 'choice' rather than 'design'.

It may help in the choice of a subject to use a list with some of the conditions that relate to human biological variation. A topic for possible study could consist of comparison of population samples that differ with respect to items in one of the rows of Table 1.1, but are the same with respect to the items in other rows.

For instance, one could study the Variations in the *gross anatomy* (1) of *the thorax* (2) at *high and low altitude* (3) in *Chinese* (4) *male* (5) *adolescents* (6), or the Comparison of the *ages of emergence through the gums* (1) of *the secondary teeth* (2) in *upper and lower class* (3) *school children* (6) of *each sex* (5) in *Managua* (4). Such studies are directed to more general problems: respectively the effect of altitude on growth of the lungs and the role of social environment in dental development. By specifying the conditions one can assess the probable usefulness of a proposed study relative to the state of knowledge from past studies using similar or different combinations of factors. When the design shows little or no improvement over past studies and the conditions closely duplicate those of the past, little is gained unless the results of the prior studies are contradictory, biased in some ways, or samples were too small. If the purpose of a study is to check on past work, some overlap of the conditions is desirable, but it is also well to explore further relationships by including some additional variables. Originality of a study usually lies in the exploration of more relationships of some phenomenon than had been successfully demonstrated by prior work. Scientific breakthrough may need insight, but the day-to-day work of biological anthropology proceeds through expanding the comparisons to different combinations of conditions. This in turn may disclose larger questions that permit new insights; that is, even small accretions of fact may suggest different ways of studying the larger subject and may raise questions of general theory.

Since the purpose of research is to acquire new knowledge, some aspects of research activity cannot be planned. One learns as one works and what one learns should influence one's further work. Although forethought is almost always helpful, at some point getting on with the job must supplant planning. In anthropological work there are often more imponderables than in other types of research. Field work, survey work, and especially longitudinal, prospective and to-be-followed-up studies need to be planned so that some aspects are consistently pursued from the start, but other aspects are flexible and subject to repeated rethinking as the study progresses and knowledge is attained.

The setting

Although more limited objectives are legitimate, a primary purpose of human biology research is to learn something about human beings in general. In many ways any subpopulation of human beings is more or less representative of the whole species. A purpose of comparative studies is to delimit the range of variability. Such studies measure the degrees of similarities and differences. Biological variables express different fractions of their variability within and between subpopulations. The less the relative variation between subpopulations, the greater the likelihood that other subpopulations are also similar and that one can generalize results to the whole population. Furthermore, as comparisons are made among more groups and the groups become more representative, general conclusions gain credibility.

For example, a study by Hulse (1964) has been interpreted to indicate that when inbred human groups breed out, growth in stature and some other body dimensions increases. Hulse examined Italian-speaking Swiss from the canton of Ticino (and most of them from two alpine valleys, Val Maggia and Val Verasca). The offspring of marriages of parents from the same village were assumed to be appreciably more inbred than those of parents who came from two different villages. Hulse studied individuals in two environments (Switzerland and California) and took account of differences in mean age between samples. Other studies in different settings gave different results, however. This tends to limit any conclusion about an outbreeding effect in children of village-endogamous marriages to situations where there was considerable inbreeding to start with, since studies of small or less inbred samples may demonstrate little or no signicant differences (Lasker *et al.*, 1990).

The studies of concern in this book refer to a single species that encompasses all living and recent human beings. The general similarity to each other of human beings allows for a choice of settings based on many factors: Besides appropriateness for the purpose of the study, one considers one's command of the language of the subjects, prior acquaintance with local individuals or with others who can provide useful contacts, conditions of living for the researcher at the site of the research, and probable costs of working there relative to available financial resources. In selecting a site for study one should not neglect the question of safety and comfort of the field workers. Howell (1990) provides useful information on the matter of safety based on the experiences of many field workers.

Research skills

The most rigidly set precondition of a study is the technical competence of the persons responsible for it. One should plan to do work for which one has learned the necessary competences or can do so in advance. It is essential to have practiced procedures in the laboratory before going into the field. If one has not taken appropriate laboratory courses in the university or had other training by competent mentors, it helps to visit such people to profit by their experience. Most scientists are willing to help a visitor learn a specific field procedure. Many problems are studied

---

BOX 1.1.   Choosing field research projects

Bernard (1988) discusses different aspects of choosing a research project. Although he deals with field work in social anthropology, his five points are also applicable to biological anthropology.

1. *You must have an interest in your problem and in the people who are your subjects.* This seems self evident. It takes a lot of motivation to put up with the hardships usually encountered in the field, and the financial rewards of such a career are usually meagre.

2. *The problem must be one susceptible to scientific investigation.* Can it be couched in operational terms? That is, if I do this and find that can I infer so-and-so?

3. *The scope of the problem must be such that you can accomplish it within the time and with the available financial resources.* I have some reservation about the application of this point to human biology. A small preliminary study may set the problem, determine the best methods and even begin to show some results. Foundations that support research want to be assured that the researcher can do what he or she says and see it through to a conclusion. A pilot study may well give such assurance.

4. *The project must be ethical.* Just saying so will carry little weight. One's ethical standards come from the culture of one's own society. The researcher should be familiar with applicable laws and be aware that violation of the ethical standards of the profession can meet with severe sanctions: isolation and even ostracism.

5. *The problem should be of some theoretical interest.* That is, the results should be applicable to situations beyond the ones tested in the study itself. Different researchers vary in their opinions as to how far speculation should go, however.

Reference
Bernard, H. Russell (1988). *Research Methods in Cultural Anthropology.* Newbury Park, California: Sage Publications.

---

through collaboration and useful collaboration sometimes gets started through a visit to a laboratory or office of someone working on a related project.

During the training one can repeat and check results and evaluate reliability. A pilot study may be possible. Preliminary studies of reliability should be systematic enough to warrant reporting the results. It is desirable to continue to duplicate some procedures on the same individuals during a study not only to minimize technical error but also to measure the probable extent of unavoidable remaining error.

In the past a field worker often worked alone or with trained local assistants or university students. In today's circumstances, most large studies involve teams in which competencies of members complement each other. The work is distributed accordingly. The participants in a joint research project should not only fulfil their assigned tasks but should try to understand and be helpful to the other members of the team because authors of research reports are properly any of those who contributed importantly to any part of the planning, execution or interpretation of the research, but all authors are responsible for all that is said in a published report.

From the point of view of sponsors, training and retraining constitute one product of research. Doing research with others and ready communication among participants are considered good training. Such training is important because poor work cannot be redeemed later. For instance, Bolsden and Jeune (1990) concluded from the reports of women on their age at menopause that the answers were warped by digit preference (reporting age to an age ending in '0' or '5'). They adjusted for this statistically on the assumption that the reported age was the nearest round number and that the actual pattern of ages was continuous. Nevertheless, conclusions from such studies remain in some doubt until the studies are repeated with more exact evidence about ages.

An example of controversy stemming from incorrect data is seen in the 1990 US Census of Population. There were many omissions in the counts in large cities and debate is still heated over whether and to what extent subsequent statistical adjustments to the counts are justifiable. Even after more careful repetition, uncertainties about the extent of previous errors can plague the interpretation. One should strive for accuracy from the beginning.

We all have shortcomings in knowledge and skills. It is easier to acquire some knowledge than to develop most skills, so it is often better to take responsibility for aspects of research for which one has the skills to proceed, than for aspects about which one knows a great deal but for which one has not yet mastered the technical skills. However, both methods and knowledge are constantly advancing so there is need both to 'retool' to apply the techniques being developed and to keep up on the literature. Through practising and reading one always remains a student.

Computer literacy is one of the skills necessary for field and survey research. It may be feasible to use a portable computer in the field. Some instruments, including anthropometric measuring devices, can feed original data directly into a computer file. Even when that is not possible, well-planned studies use record forms that permit transfer of the data to computer files with minimum risk of errors. Letters of the Roman alphabet and Arabic numerals are the symbols least likely to lead to later ambiguities. Accents and other double strike symbols are the most likely to lead to later problems. Of course at the analysis stage of the work it may be possible to collaborate with a person who has greater command of programming and programs, but it is wise to plan for this in advance so the data will be in ready-to-use form.

Risk of bias

The matched controls possible in experimental work on other species usually can be only roughly approximated in field work studies in human biology. The strategy therefore must be to make comparisons between groups within the population that differ from each other with respect to a critical variable. Samples for such comparisons must be selected in a way that is unbiased with respect to any covariable that may be involved, or, if there is a bias, one that does not favor the hypothesis being tested.

With respect to the comparison of stature between American-born and immigrant Chinese which was the basis for my PhD dissertation and is cited in Box 1.2, the bias does prevent the simple interpretation of the results. From the prior studies of Boas (see Chapter 3 and Chapter 4 (Box 4.1)) and a few others (cited by Kaplan, 1954), it was possible to state the hypothesis that adult stature depends to some extent on the conditions of life during childhood growth. Did the conditions during their rearing in the United States of Chinese born there lead to greater stature than that of first-generation immigrants from China? However, it was already known that adult stature gradually had been increasing in most parts of the world (presumably including China). It was also known that people shrink in stature at older ages because of compression of their cartilaginous intervertebral discs and changes in their posture. Thus older Chinese men anywhere would be expected to have shorter stature, on average, than younger ones. The finding of greater stature in the American-born than in the immigrant Chinese who were mostly older, is therefore subject to alternative explanations. Since the samples were matched in most respects (including the counties in China where their ancestor lived) the greater average stature of the American-born compared with the immigrant Chinese might be due to the different environments in which they grew up, as hypothesized, but it also might be explained by a bias that had not been excluded: on average, the American-born were considerably younger than the immigrants. It was only because of other evidence that adult stature changes little (e.g., Roche, 1992, Fig. 4.1) that it could be argued, despite shortcomings in the design of the study, that the differences in the circumstances of growing up in China and in the United States did have a substantial influence on the stature of Chinese men. The only bias that a good study can tolerate is one that disfavors the hypothesis.

In observations about dental decay in the same thesis research project (Lasker, 1945), the bias was against the hypothesis. The etiology of dental caries was less well understood then than now, and genetic influences were exaggerated in some publications about the causes. The idea of a predominant environmental influence (diet) was sometimes underrated and the added evidence of an increase of this dental disease in Chinese born in America therefore was of interest. But could the finding have been biased by the age difference? No. Quite the contrary. There were more carious, filled and lost (i.e. extracted) teeth in the American-born than in the Chinese immigrants. Dental caries is progressive. Therefore, after the permanent dentition has emerged through the gums, the number of diseased, filled and missing teeth can only increase. The finding of more caries in the younger American-born than in the older immigrants implies that growing up in America was related to the observed difference and that the study design underestimated the difference. That is, Chinese subjects of the *same* age might be expected to show an even greater influence of the diet and other factors associated with the life in America on those born there. Subsequent studies of other populations have shown that great use of sugar and other refined carbohydrates had a profound effect on dental caries, especially before the widespread use of fluorides in the prevention of dental caries.

When these studies were being done, multivariate methods of statistical analysis such as multiple regression were known, but the burden of calculations on a calculating machine of the time were onerous. Although it was then impractical, today anthropologists deal with a problem such as a disparity in ages by using multivariate statistics to control for age or other differences.

One should bear in mind, however, that control by multiple regression does not completely solve the problem. It is best to record the exact date of birth and to calculate age as a decimal fraction of years. This is often not possible in field situations, however. Then, even if the regression on age is linear (or converted to linear by some formula) regression on stated or estimated age makes incomplete allowance for age.

Suppose one allows for age in a study of body size and achievement in school children aged 5 to 10 years old. Then one might include in the same category children who were just barely 6 and others aged 6 years and 11 months. But the latter would be likely to be larger and know more than

BOX 1.2. Learning by error

In 1939 I decided to do for my doctor's thesis a study using Franz Boas' strategy (see Chapter 3 and Chapter 4). At that time most anthropologists believed that anthropometric characteristics were racial and fixed by heredity. Boas had found that American-born sons of immigrants had statures and cephalic indices (c.i.) different on average from those of their fathers. His interpretation that this was caused by environmental factors was still being challenged, however, and I planned to compare measurements and observations of first-generation immigrant Cantonese Chinese with those born in America.

My Professor at Harvard, E. A. Hooton, was not attracted by the hypothesis of a significant environmental component in these traits, but he let his students explore for themselves and he told me: 'Go ahead – and while you are doing it measure as many Chinese as possible from every part of China so you can describe the regional types.'

I also went to see Boas at Columbia University. He spoke about his experience in studying the cephalic index and said: 'what will it take to demonstrate a statistically significant difference? Suppose the standard deviation is about 4 [c.i.] units and the difference between immigrants and American-born averages 1 unit; then, let me see, three probable errors of difference [equals about 2 standard errors] would usually require samples of more than 100 immigrants and 100 American-born to show a significant difference at $p = 0.05$. You will find it hard to measure that many.' By the logic of Boas' power analysis (see Chapter 2) it would have required larger samples than I was able to measure, and the difference I found in c.i. was not statistically significant – although some other differences were.

The final thesis was about 3 inches thick. Hooton, with his usual good nature, bellowed: 'The biggest goddam thesis I ever saw; heh, heh; really two theses; heh, heh; The one I wanted you to write and the one you wanted to write yourself; heh, heh, heh.'

Except for a piece about the dentition, only the comparison of immigrants with American-born ever appeared in print (Lasker, 1946). The experience taught me that one has to start somewhere and even if one's hypothesis is not fully sustained, some of the results may be of interest. Knowledge is cumulative so that studies that are too small to yield significant results in one respect may be large enough to answer another question or to serve as a pilot for further studies.

Reference
Lasker, G. W. (1946). Migration and physical differentiation. *American Journal of Physical Anthropology*, ns 4, 273–300.

the former and the allowance for age would be incomplete. This would be even more likely if some of the recorded ages are erroneous. A child who is said to be, and therefore mistakenly recorded as, 6 years old but who is actually 7 is very much more likely to be larger and know more than another child who is 5 years old but said to be and recorded as 6. Thus within each age category the age-related variations in the two variables would still lead to an apparent positive correlation between size and knowledge 'after allowing for age' because the allowance always tends to be incomplete. In a study such as that of the adult Chinese, where the sole source of information about age was self-report of the subjects, problems of this kind may be considerable and the only safeguards are care in the initial collection of the data and, insofar as possible, matching the subjects in the different study groups in respect to age and other confounding variables.

If unavoidable errors creep in they may tend to affect within-group regressions as explained above, but if such errors are similar in number, direction and extent in various groups, they should have little effect on differences between groups. This stricture not only applies to age, of course; statistical allowance for social status is also susceptible to the problem because occupation, wealth and education only partially account for the underlying standard of living and such variables are particularly likely to be recorded in crude categories or with unavoidable errors. Even sophisticated statistical studies allowing for such factors as geography by regressions against distance are subject to the same problem because, however measured, distance is an incomplete measure of the biologically significant behavior that is being allowed for.

Grants

Planning for research is a step in doing research. A usual product of a research plan is a research grant application or proposal. Such a proposal normally has all the elements of a research report except that it does not say whether the hypothesized result is so or is not so. In anthropology, however, and especially in field studies, it is often desirable to structure research plans loosely so that conditions not anticipated in the plan – and sometimes wholly or in part new – can be included when opportunity affords. This is one of the reasons why biological anthropologists often find themselves at a disadvantage when applying to granting committees that also handle applications from 'hard' sciences or when the committee in turn reports to such an agency. One should state any such problem explicitly because the difference from other sciences is only one of degree; all research workers modify their research questions during the

BOX 1.3.   Does planning account for advances?

It would not be appropriate to present the case for planning without the counterargument that science is the testing of ideas: the daughter of circumstance and insight. Some experiences with tests that were not planned illustrate the point.

In 1948, together with F. G. Evans and B. A. Kaplan, I planned a study of the effects of living in the United States on bodily measurements of Mexicans by studying in Mexico returned emigrants and brothers and sisters who had not emigrated. Since we were not sure what conditions would be found to be important and in order to explore other questions, we collected a variety of kinds of information. In the event, there were not enough brothers and sisters of different categories for a useful test of siblings as planned, but we found that individuals who had lived more of their young lives in the United States were taller and larger in some other ways than those who had migrated when older or not at all (Lasker, 1952; Lasker and Evans, 1961). Some of the items of information collected during the field work were not utilized in the initial study. In the years since, however, they have been found to apply to a variety of questions, some of which arose only because of subsequent research.

For instance, in the course of analysis it became clear that most of the measurements varied with age, and the ages of those in different migrant and sedente categories differed. To measure the age effect required larger samples. New data were not collected for this purpose, but data from Goldstein's (1943) and our study were reanalyzed to study aging (Lasker, 1953).

Although information on ethnicity was not utilized in the analysis of migration effects, a question about it has been included. In translation it asked: 'What were your father and your mother, Indian or Mestizo?' The eventual analysis of responses, not part of the original plan of study, showed that responses were strongly related to the age and social status of the respondent, but we did not find consistent differences in physical characteristics (Kaplan and Lasker, 1953).

One of the issues in interpreting migration studies is the question of whether the migrants change their growth pattern following migration or whether they were different from the outset (were selected). This issue was the reason for the original plan to study siblings, but only a subsequent visit to the site, not planned at the outset, permitted a prospective study of possible selective migration. Those subjects who had been to the United States within the four years *after* they were measured were not significantly different anthropometrically from those who had not, thus there was scant evidence of physical selection during those four years (Lasker, 1954).

Data from the original Mexican work were subsequently also mined for previously unplanned studies of migration, isolation, genetic drift, natural selection through fertility and mortality, isonymy and other topics. The chief weakness of the articles about these applications is the limited sample sizes and not the omission from the original plan.

Most recently we compared the offspring of endogamous and exogamous matings (Lasker *et al.*, 1990); again, use of the data collected for other purposes is a useful strategy for trying to estimate influences of this scale.

References

Goldstein, M. S. (1943). *Demographic and Bodily Changes in Descendants of Mexican Immigrants with Comparable Data on Parents and Children in Mexico.* Austin: Institute of Latin American Studies, University of Texas.

Kaplan, B. A. & Lasker, G. W. (1953). Ethnic identification in an Indian Mestizo community. *Phylon*, **14**, 179–90.

Lasker, G. W. (1952). Environmental growth factors and selective migration. *Human Biology*, **24**, 262–89.

Lasker, G. W. (1953). The age factor in bodily measurements of adult male and female Mexicans. *Human Biology*, **25**, 50–63.

Lasker, G. W. (1954). The question of physical selection of Mexican migrants to the U.S.A. *Human Biology*, **29**, 52–8.

Lasker, G. W. & Evans, F. G. (1961). Age, environment and migration: Further anthropometric findings on migrant and non-migrant Mexicans. *American Journal of Physical Anthropology*, **19**, 203–11.

Lasker, G. W., Kaplan, B. A. & Sedensky, J. A. (1990). Are there anthropometric differences between the offspring of endogamous and exogamous matings? *Human Biology*, **62**, 247–9.

research to take into account what they have already found. The complaint is sometimes made that those who give (or even say 'invest') funds for research want to be sure of the result and feel this is only likely if all procedures to be followed and alternative possible outcomes are spelled out in a grant proposal that has the form of a research report. Therefore, if one applies for support, even to a college committee, one must assume that the scientists on it will understand a stated need to retain some flexibility. One can outline a research grant request in a way similar to a report. Effective applications always include information about planned sample sizes and duration of the study so that potential donors can judge whether one is likely to be able to produce results of value.

Fortunately there are some foundations, universities and other sponsors of research with experience about, and sympathy for, biological anthropology, and their staffs understand our research strategies. Most such institutions view it as part of their missions to provide opportunities for professional development through the research activities they support. Nevertheless, the researcher's past record also carries weight. That is why successful completion of prior work and its presentation at professional meetings and through publication is viewed as preparation for the next piece of research. Even students should complete small projects as part of their preparation for larger projects and for a career in research. This may be even more important than the accumulation of knowledge about research findings. Past experience in research is more important than any advice that can be given, but advice to think consciously and plan thoroughly should not be ignored.

A model grant application

In setting forth a plan most reseachers go through exactly the same steps they anticipate repeating in carrying it out and publishing it. These are conventionally divided into 1. Introduction, 2. Sample, 3. Methods, 4. Results (but see point 4 below), 5. Discussion, 6. Abstract, and 7. Literature cited.

1. Introduction

The first task is to define the *problem*. State the hypothesis to be tested. Save consideration of side issues for the discussion. Most references based on reading notes belong in the discussion section rather than in the introduction.

2. Subjects

The plan for the selection of human subjects is critical. They must be a random sample of some larger 'population' in order for the results to have any general applicability. For a statement to apply generally to a population it should be based on statistically significant results on a representative sample. For instance, such a representative sample of adult Japanese would be one in which any adult Japanese person is equally likely to be included. Since such a random sample is not always practicable, an alternative strategy is to draw stratified samples in which subsamples by age, sex and any other variable that might influence the traits under study are separately considered and the results for subsamples are weighted by the numbers of members with the respective characteristics in the whole Japanese national population. That is, the data on Japanese of every sex–age–occupational–regional group is multiplied by the number of members of that group in the population and the results are divided by the sum of the numbers of members of all groups to yield weighted statistics for the whole population.

When selecting a sample, applying grounds for exclusion of individuals such as 'those who look mixed', 'those with foreign names', 'the elderly', etc. may leave samples that are difficult to place relative to the population and this is particularly true if the grounds for exclusion are ill-defined.

Random samples may be drawn from administrative or other lists or from a census of habitations. In some genetic studies relatives of other subjects are excluded; any such practice must be stated, however, because it renders the sample one of families rather than of individuals in the population. In general, the broader the population, the more likely the results will have some validity for the whole human species, but the broader the definition of the population, the more difficult to sample it at random. Thus the kinds of sample drawn depend on the nature of the problem. Those with limited experience sometimes try for too ambitious an objective when a study of more limited scope about a narrowly defined population would have been more appropriate considering available resources.

One must plan how to seek consent to study and how to deal with individuals who do not wish to continue to cooperate. This happens with long schedules or intrusive procedures such as the collection of blood specimens. Refusals may interfere with a carefully laid plan for randomizing the sample. Therefore it is important to plan to gain some information about those individuals whom one wished to include but who did not volunteer to do so. The plan should sequence the acquisition of information. After receiving permission to proceed, one asks about age, social categories, and other matters that are perceived by the respondents as harmless. The answers to these questions are recorded before any attempt to deal with more sensitive aspects of the protocol. In that way the subsample on which only some items are completed can be compared with the rest of the sample to assess the representativeness of the subsample that provides information on the sensitive items. If the two subsamples are not significantly different in those variables for which one has information on both, one can have more confidence that the more sensitive items can be generalized to the whole sample and hence, if the sample was random, to the population as a whole.

It is important to plan the sampling method carefully so that it can be maintained throughout the study. A change in the criteria for sampling during a study may make it impossible to interpret how the results relate to the population. Sometimes it is desirable or even necessary to determine the size of the sample in advance. For some suggestions for estimating minimum sample sizes needed see Chapter 2.

3. Methods

Methods also must be systematically applied, hence carefully planned. Quite apart from the fact that a researcher supported by a grant may be obliged to follow the terms of the grant, any change in midstudy may

BOX 1.4. Abbreviated project summary

An example with weaknesses as well as strengths illustrates problems in grant applications. The 1980 project described here was subsequently modified and the submissions to two private foundations were successful.

1. Title. 'Structure of Human Population of Britain Studied by Surname Analysis.'
    Weakness. Overseas work generally has a low priority with US federal granting agencies; however, in anthropology this does not apply to out-of-the-way places, but Britain would not be so considered.
    Strengths. Surname analyses involve large numbers of names and require huge samples. It is precisely in cosmopolitan populations such as that of Britain where surnames are diverse and large samples are readily available.

2. Studies of local populations in England showed that the breeding structure is well represented by the geographic distribution of surnames. The coefficient of relationship by isonymy (Ri = half the sum of products of the frequencies of each surname in two samples of names) will be applied.
    Weaknesses. The method is not new. Surnames are an indirect way to approach the problem of population structure. Structure in developed societies is thought to have little genetic effect since inbreeding may be important in isolates but is limited in the general population of such societies.
    Strength. Although the method is most appropriate for large populations like that of Britain, the prior applications of the method have been to small communities.

3. Registers of marriage in England and Wales for January–March 1975 list 165 487 entries and 29 829 different surnames. The incidences will be analyzed according to ethnicity and locality.
    Weakness. Determining ethnicity from surnames is subject to errors.
    Strength. The data are adequate for the main purposes.

4. Values of Ri will be calculated and analyzed 1. between pairs of registration districts, 2. with other sets of data, and 3. for the effects of distance.
    Strength. These objectives are concrete and achievable.

5. Historical data on a 2% sample of England and Wales in 1851 will be compared with the 1975 data.
    Weaknesses. Scientists are often little interested in historical, sociological and humanistic aspects. The 2% sample was to be stratified, not random. [The 2% sample was never completed and this point was deleted from the revised application.]

> *Strength.* A study at one point in time gives only one kind of information, that on status at the time. Addition of another time period gives two more kinds of information, that on status at the second time, and that about changes over time.
> 6. Collaboration will be fostered with ten other scholars (listed).
> *Strength.* Co-investigators are a plus.
> *Weakness.* Listing distinguished scholars as co-investigators disqualifies them as references or reviewers.
> 7. Personnel and dissemination of results (listed).
> *Strength.* The investigator has a 'track record' of published results.
> *Weakness.* Experienced investigators cannot claim 'development of personnel' as a secondary goal unless they involve graduate students and postdocs in the work.

make it impossible to compare data collected early in the work with those collected later. Nevertheless, one can draw subsamples for study by unanticipated procedures; if the original methods are continued at the same time, it should be possible to draw some inferences about the whole sample from results of special tests on a subsample, provided that the subsample is comparable to the rest of the sample. Thus it is well to anticipate what methods will be used, but allow for the possibility of adding improved methods to the protocol if they become available. Use of a subsample may also be necessary for very expensive or time-consuming methods. For instance, human body fat can be assayed by many different methods (Shephard, 1991), some of which, such as calliper measurements of skinfolds, are simple enough to apply to large numbers of individuals in the field. It may also be possible to bring a subsample of the subjects into a laboratory for ultrasound, bioelectric impedance or underwater weighing studies.

4. Results
This is the one step of research than one can not, or at least should not, be able to plan. Results that are pre-planned are of no interest. Even an elementary classroom exercise in science can be of more interest if the outcome is not completely pre-planned. The reward of research is learning something that was not previously known, even if the something that was not known is only a small refinement or extension to a slightly different situation of some well-established fact. Too great certainty in advance has tempted some to overinterpret or misrepresent their data. If one were certain of the result, the study would not be worth doing. On the other hand, it takes a certain confidence about the possibility of a result to

be sufficiently motivated to invest adequate energy in a study. What is needed is enthusiasm for the possibility of a result, but reserved judgment. Reservations based on critical consideration of accepted facts (from one's own work or that of others) open up hypotheses for further research.

5. Discussion

Although it is sometimes done in the introduction, the discussion is the best place in a grant proposal to bring up literature about research with different material, methods or ends. A search of related topics in other sciences can help define the problem in broader terms. Some researchers tend to get lost at this stage, however. There comes a time to put the journals aside and to begin to look for oneself.

6. Abstract

The project summary is usually placed at the beginning of the research proposal, but, since it deals with every other part, it is the last part to prepare. One should write the kind of abstract that summarizes, not the kind that merely says what the project is about. If one thinks one knows what the outcome of the work will be, do not make a mystery of it. State it as a hypothesis from the outset. The results of a study cannot be known at the time of writing the plan, however, so the abstract will be about the hypotheses, material and methods, not about results.

7. Literature cited

The list of references is one part of a final report that can and should be practically completed while the work is being planned. Unless one knows the present state of knowledge on a subject, it is difficult to justify doing any research. The librarian can usually help get one started with a literature search and the *Science Citation Index* can help one from there.

References

Bolsden, J. L. & B. Jeune (1990). Distribution of age at menopause in two Danish samples. *Human Biology*, **62**, 291–300.

Harrison, G. A., Tanner, J. M., Pilbeam, D. R. & Baker, P. T. (1988). *Human Biology*. 3rd edn. Oxford: Oxford University Press.

Howell, Nancy (1990). *Surviving Fieldwork*. Washington DC: American Anthropological Association.

Hulse, F. S. (1964). Exogamy and heterosis (with a comment by G. W. Lasker). *Yearbook of Physical Anthropogy*, **9**, 240–57. (Translated into English from Exogamie et hétérosis. *Archives Suisse d'Anthropologie Général*, **22**, 103–25, 1957.)

Kaplan, B. A. (1954). Environment and Human Plasticity. *American Anthropologist*, **56**, 780–800.

Lasker, G. W. (1945). Observations of the teeth of Chinese born and reared in China and America. *American Journal of Physical Anthropology*, n.s. **3**, 129–50.

Lasker, G. W., Kaplan, B. A. & Sedensky J. A. (1990). Are there differences between the offspring of endogamous and exogamous matings? *Human Biology*, **62**, 247–9.

Parkes, Alan S. (1985). *Off-beat Biologist.* Cambridge: The Galton Foundation.

Roche, Alex F. (1992). *Growth, Maturation and Body Composition: The Fels Longitudinal Study 1929–1991.* Cambridge: Cambridge University Press.

Shephard, Roy J. (1991). *Body Composition in Biological Anthropology.* Cambridge: Cambridge University Press.

2 Research designs and sampling strategies

C. G. N. MASCIE-TAYLOR

Some people have latched on to the importance of statistics and others not. Over a hundred years ago, Francis Galton (1889) wrote 'some people hate the very name of statistics but I find them full of beauty and interest. Whenever they are not brutalised, but delicately handled by the higher methods, and are warily interpreted, their power of dealing with complicated phenomena is extraordinary. They are the only tools by which an opening can be cut through the formidable thicket of difficulties that bars the path of those who pursue the Science of man.'

The failure to use statistical methods when they ought to be used is a fairly common occurrence. Gore and Altman (1982) quote a divorce case in 1949 in which the sole evidence of adultery was that a baby was born almost 50 weeks after the husband had gone abroad on military service. The appeal judges agreed that the limit of credibility had to be drawn somewhere, but on medical evidence 349 days, whilst improbable, was scientifically possible. So the case failed. Was the husband hard done by? Saying that an event is possible is quite different from saying that it has a probability. In this case the probability of a pregnancy of 349 days was of the order of one in 100 000.

Sackett (1979) identified 56 possible biases that may arise in analytic research, over two thirds of which related to study design and data collections. Errors in data analysis or interpretation can usually be rectified but deficiencies in design are nearly always irremediable. Schor and Karten (1966) undertook a statistical evaluation of 149 medical journal manuscripts. Only 28% were judged acceptable, 67% were deficient but could be improved and 5% were totally unsalvageable.

Designing a study

The way in which a study is designed is therefore of major importance. Since it is nearly always impossible to measure everybody, due to some logistical constraint, the common practice is to obtain a sample of the total population. Thus research design includes (a) designation of the population of reference (b) definition of the type of study sample (c) size of sample and (d) type of statistical design.

The type of research design used is either experimental or observational. In the observational approach the investigator has to observe natural outcomes resulting from differing exposures. No manipulation is possible. For example, people differ in blood groups. Thus one can observe whether people with different blood groups (e.g. blood group A versus blood group O of the ABO system) vary in their susceptibility to some disease. It is commonly found that people of blood group A are more likely to develop carcinoma of the stomach than are individuals of blood group O (Aird and Bentall, 1953). The problem (see section on observational studies below) with this observational approach is that the observed groups may differ in ways other than the variable (attribute) in question.

In experimental studies the investigator has control of some feature that when varied may be associated with different outcomes. For instance in animal studies the investigator may control the diet and relate changes in nutritional intake to reproductive success and growth. In human studies, however, ethical considerations limit the applicability of the experimental approach and the investigator often has to use the observational approach.

Even so there are a number of experimental designs that are commonly used by anthropologists, physiologists and related medical disciplines. This chapter briefly reviews the advantages and disadvantages of the five main categories of experimental design.

The randomised control trial

The strongest research design is the *randomised control trial* (RCT) whereby subjects are allocated randomly to two groups, an intervention group and a control group. This approach is commonly used in clinical trials for new treatments and the design is also called *randomised clinical trials*. It is important to distinguish between random selection of subjects from a population, which can occur in both observational and experimental designs, and the subsequent random allocation of subjects into intervention and non-intervention groups that only occurs in the experimental approach.

The impact of helminth infections on human nutrition and growth is one example of a group commonly tested by the RCT design which randomly allocates subjects to treatment and placebo groups. The experimental group receives the new treatment, whereas the control group gets nothing, conventional therapy, or a placebo. Ideally, in order to minimise various biases, neither the patient nor the researcher knows who got what until after the trial is over. This is called a 'double blind' study.

The major advantage of the RCT is that it ensures that there are no systemtic differences between the groups. Since they are drawn at random from the same population, randomisation ensures that they will be identical except for differences that might arise by chance. There are potential drawbacks. For instance the trials designed to look at the effects of aspirin on reinfarction rate after heart attacks cost up to £4 million pounds ($7 millions) while the study designed to look at lipids and infarction cost £90 million pounds ($150 millions)! The other drawback is that subjects who agree to participate in an RCT may not be representative of subjects in general.

There is, of course, an ethical dimension to such clinical trials, which is that experimentation is based on a judgment that in certain circumstances it is legitimate to put a subject at possibly greater risk, with his or her consent, because of the overriding need of society for progress in combating certain diseases.

Stephenson (1987) gives four reasons why an ethically constructed placebo group is essential if the aim is to determine just how much malnutrition due to helminth infection can be alleviated by treatment. The reasons are 1. a placebo group can be ethically included in many studies of chronic helminthiases of unclear pathogenic potential because treatment is not withheld but is delayed only until the end of the study, 2. in many areas where helminths are endemic, the treatment that subjects receive reaches them sooner than the overcrowded and understaffed conventional health facilities could provide, 3. all study subjects benefit from participation, regardless of whether they are initially given a placebo or not, and 4. only with the inclusion of a placebo group can one estimate the magnitude of seasonal, placebo or unpredictable effects on changes in malnutrition and parasitism over the duration of the study.

The importance of including a placebo group can be seen from the results presented in Table 2.1. They show the increases in haemoglobin and three anthropometric measurements in placebo and treated (using the drug metrifonate) groups of Kenyan children with *Schistosoma haematobium* infections 6 months after treatment. Failure to include a placebo group would have (a) over-estimated the haemoglobin rise by three-fold in treated children, and (b) under-estimated the improvements in growth due to treatment with metrifonate.

This study also illustrates that a simple design is quite often the best design. This design involves only two groups (treatment and placebo) and two measurement periods – baseline or pre-treatment and post-treatment, some months later. Additional study groups and measurement periods are (a) more expensive, (b) more time consuming, and (c)

Table 2.1. *Increases in haemoglobin level and anthropometric measurements in placebo and metrifonate-treated groups of Kenyan children with* S. haematobium *infections showing need for inclusion of placebo group*

Parameter	Group*	Increase 6 months after treatment	*t*-test *P*
Haemoglobin	Placebo	1.0 ± 0.8	0.014
(g/dl)	Metrifonate	1.3 ± 0.8	
% weight for	Placebo	−0.5 ± 0.22	0.001
age	Metrifonate	1.8 ± 0.19	
% arm circumference	Placebo	−1.3 ± 0.19	0.001
for age	Metrifonate	0.4 ± 0.17	
% triceps skinfold	Placebo	−1.4 ± 0.67	0.001
thickness for age	Metrifonate	7.7 ± 0.58	

*Sample sizes: placebo = 198; metrifonate = 202 (haemoglobin), 201 (anthropometry).
Adapted from Stephenson *et al.* (1985a,b).

more complicated to analyse statistically. However, testing two drugs with one placebo control may be less expensive than tests of each drug against its own control. Another refinement which is commonly included is the design of taking the treated group off treatment and putting the controls on the treatment to see if the former fall back and the latter repeat the performance of the first treatment group.

The cross-sectional survey

The deworming study just described is an example of a longitudinal rather than a cross-sectional study (at one point in time). Cross-sectional studies are relatively cheap and subjects are neither deliberately exposed to possibly harmful agents nor do they have treatments withheld from, or imposed on, them. However, cross-sectional studies have a major limitation – it may not be possible to state what is cause and what is effect.

For example, if a cross-sectional study had been conducted between *S. haematobium* infection and haemoglobin status, which revealed an association between infection and haemoglobin status, it would not be possible to prove that the infection caused the haemoglobin level. Secondly, cross-sectional studies are likely to produce false negative results because different subjects in a community may contract the helminth infections at various times throughout the year. Thus a study at one time of year may not apply well to what happens at other seasons.

The cohort study

One alternative to the RCT is the cohort study. Two groups (or cohorts) are identified, one of which has been exposed to the putative causal agent and another which has not. Although good cohort studies are not expensive, the major disadvantage is that it is impossible to be sure that the groups are comparable in terms of other factors which may influence the results. Smoking, for example, is related to social class and jobs that involve exposure to carcinogens are usually related to social class as well. So if smokers have a higher incidence of lung cancer than non-smokers, to what extent is lung cancer attributable to the smoking, and to what extent to differences in social class, occupational exposure or something else?

A second drawback is that untreated controls may be difficult to find in certain circumstances. Lastly, it may be difficult, if not impossible, to maintain blindness; for example, subjects would be aware of who smoked and who did not.

The retrospective (case-control) study

In this design, a group of people who already have the outcome (cases) is matched with a group who do not (controls), and the researcher determines the proportion in each who had previously been exposed to the suspected causal factor. For instance, people have some attribute such as blood group A, or exposure, such as contraceptive usage, and the development of disease in the group with the attribute is compared with that in the group without. Because the design is retrospective it cannot account for other factors which could lead to the outcome. Thus the observed groups may differ in other ways in addition to the attribute in question, thereby confounding the comparison.

Case-control studies have been attempted with presence or absence of a specified helminth infection as the criterion for selection. However, there is usually a significant relationship between socio-economic status and prevalence and intensity of helminth infections such that the relatively wealthy people do not get intestinal parasitic infections whereas the urban and rural poor have them for much of their lives. Thus a study of round worm (*Ascaris*) infected children in Panama revealed that they came from low income homes and had mothers less educated than did their uninfected counterparts.

Another example is the test of the hypothesis that smoking is associated with lung cancer. Suppose 200 cases of lung cancer are matched with 200 controls free of lung cancer from the general population (matched by age, sex and socio-economic status). Then smoking habits of the two groups are ascertained. The results shown in Box 2.1 clearly indicate a

---

BOX 2.1. Testing a hypothesis and the odds ratio

The hypothesis is that smoking is associated with lung cancer. To test the hypothesis take 200 cases of lung cancer and then select 200 controls free of lung cancer from the general population but matched by age, sex and socio-economic status. Then determine the smoking habits of members of the two groups. Suppose the results are as shown below:

	Cases	Controls
Smokers	185 (*a*)	75 (*b*)
Nonsmokers	15 (*c*)	125 (*d*)
Total	200	200

$\chi^2 = 132.97$ $p < 0.001$. There is a highly significant association between smoking and lung cancer.

$$\text{odds ratio} = \frac{a \times d}{b \times c} = \frac{185 \times 125}{15 \times 75} = 20.56$$

Thus the relative risk for smokers is 20.56.

---

marked imbalance with 185 out of the 200 lung cancer cases being smokers while only 75 of the control group smoke. Simple statistical analyses using a chi-square test reveal significant differences ($\chi^2 = 132.97$ $p < 0.001$). This test does not take into account any specific matching procedures (e.g. pairwise matching) but techniques are available to do so (see Hennekens and Buring, 1987).

Because of the retrospective nature of the study it is not possible to derive the incidence for either smokers or non-smokers. However, it is possible to estimate the relative risk provided two assumptions are valid: (a) the disease has a low incidence (less than 5%) in the general population, (b) the control group is representative of the general population with respect to the frequency of exposure.

Under such circumstances a statistic called the odds ratio gives a close approximation to the relative risk (see table in Box 2.1) and in this example the odds ratio of 20.56 provides an estimate of the relative risk for smokers of developing lung cancer.

The prospective survey
The prospective survey is similar to the cohort study but only one group is chosen initially. The aim of the study is that some of the subjects will become exposed to the putative causal agent over time and others will

not. Therefore, the outcomes can be looked at in these naturally occurring groups some time in the future. The advantages and disadvantages are similar to those of the cohort study. In addition there may be the problem that there may be too few subjects in one group to allow for proper analyses.

Observational studies

In observational studies data from a sample of individuals are used, either implicitly or explicitly, to make inferences about the population of interest. For this extrapolation to be valid, it is essential that the data obtained are as representative of the population as possible (see Chapter 5). This usually entails some type of random sampling of subjects. In developed countries ready made lists often exist (e.g. electoral roll or doctor's patients lists), but they may not be accurate or up to date. Quite often there is a desire to examine certain subgroups and this information may not be available.

So the major problem of all observational studies is selection of subjects for study. In developing countries the problem of selecting a sample is made worse because registers rarely exist, and those that do are out of date and inaccurate. It is at this point that many statistical textbooks come up with lines like 'the selection of subjects must be given careful attention at the design stage, because if the sample is not representative of the population, then the results will be unreliable and of dubious worth'. In other words simple random sampling is both costly and may be impracticable but every source of possible sampling bias must be considered, eliminated if possible, and recognised as a serious limitation of conclusions to the extent that it remains.

The advantages of simple random sampling are that (a) it is simple to conceptualise, (b) it provides the probabilistic foundation of much of statistical theory, and (c) it provides a baseline to which other methods can be compared. The disadvantages of simple random sampling are that all enumeration units in the population must be identified and labelled prior to sampling. This process is potentially so expensive and time consuming that it becomes unrealistic to implement, because in practice (a) sampled individuals may be highly dispersed, and (b) certain subgroups in the population may, by chance, be totally overlooked in the sample.

Cluster sampling

What is the alternative? The World Health Organisation say that 'the most feasible approach would be to use a multistage sampling procedure'.

This procedure involves the subdivision of the population into a number of recognisable administrative subdivisions, such as districts, from which a random sample is drawn. The first stage sampling units are further subdivided into smaller administrative divisions such as villages from which another sample is selected at random. Within the village a further random sample of households is selected. Finally, the total number of families sampled should be 'such as to provide enough individuals in each age group for statistically sound conclusions to be reached'.

The multistage sampling procedure proposed by WHO is a pragmatic attempt at resolving the issue of sampling design. However, the definition of this type of sampling procedure is really 'multistage cluster sampling'. The advantage of cluster sampling, and the reasons why it is so widely used in practice, especially in nutritional and anthropometric surveys of human populations, and in sample surveys covering large geographic areas, are feasibility and economy.

Cluster sampling may be the only feasible method since the only sample frames readily available for the target population may be lists of clusters. If that is the case, it is almost never feasible, in terms of time and resources, to compile a list of individuals (or even households) for the sole purpose of conducting a survey.

The problem with cluster sampling is that generally it will not produce as precise estimates as will simple random sampling using the same total sample size. However, due to the greatly reduced cost and administrative ease, a larger cluster sample may be selected, for the same cost, than that which is possible using other sampling schemes. As a result of the larger size, a relatively high level of precision will result.

The ratio of the variance with cluster sampling to the variance with simple random sampling is called the design effect (deff) which can be measured by

$$[1 - \text{ROH}(b - 1)]$$

where ROH is the interclass correlation coefficient and b is the cluster size.

ROH ranges from $-1/(b - 1)$ to $+1$ and is a measure of the homogeneity in a cluster. Generally, populations are not well mixed and so ROH tends to be greater than zero, sometimes by a little, sometimes by a great deal, depending on the variable in question. Even a relatively small ROH can have a large effect on the variance.

As cluster size increases, ROH tends to decrease but not at the same rate as the increase in cluster size. For example, the interclass correlation coefficient for three households next door to each other might

---

BOX 2.2.   Estimating the difference between two means

A study wishes to examine the difference in energy intake at lunch between children in a school which offers a hot school lunch and children in a school which does not. Other nutrition studies have estimated the standard deviation in energy intake among school children to be 300 joules, and the study wants to make the estimate to within 100 joules of the true difference with 95% confidence. The number in each sample (n) is

$$n = \frac{z_{1-a}^2 \, (2\sigma^2)}{d^2}$$

where z_{1-a}^2 represents the number of standard errors from the mean, σ is the standard deviation and the quantity d denotes the distance, in either direction, from the population difference, $\mu_1 - \mu_2$, and may be expressed as $d = z_{1-a}^2 \sqrt{(2\sigma^2/n)}$. (The value of 1.960 can be found using standard normal deviate tables, 95% corresponds to 1.960, 97.5% to 2.240 and 99% to 2.542.)

The formula to use is

$$n = \frac{(1.960)^2 [2(300^2)]}{100^2} = 69.15$$

Thus 70 children from each school should be studied.

If the estimate were to be made to within 75 joules of the true difference with 95% confidence the equation becomes:

$$n = \frac{(1.960)^2 [2(300^2)]}{75^2} = 122.93$$

Thus 123 children from each school should be studied.

Using the previous example but with 99% confidence:

$$n = \frac{(2.542)^2 [2(300^2)]}{75^2} = 206.78$$

Thus 207 children from each school should be studied.

---

be 0.25 whereas for the entire village of say 90 households it might be only 0.10.

Suppose a multistage cluster sampling procedure is used in a developing country to examine the nutritional status of children where the size of each cluster of children surveyed is 90. Assuming a low ROH of 0.05 the design effect (deff) can be calculated as:

$$\text{deff} = [1 + 0.05(90 - 1)]$$
$$= 5.45$$

BOX 2.3.   Hypothesis testing for two population means

One-tailed test

A study is being planned to test whether asthmatic subjects have decreased activity of the enzyme glucose 6-phosphate dehydrogenase (G6PD) than non-asthmatic subjects. From a pilot study, the standard deviation of enzyme activity is estimated at 5.1 μg Hb and is assumed to be the same for both groups. The hypothesis of no such difference is to be tested at the 5% level of significance. It is desired to have 80% power ($\beta = 0.20$) of detecting a decrease of 2 μg Hb in the asthmatic group.

The formula is

$$n = \frac{2\sigma^2(z_{1-\alpha} + z_{1-\beta 2})}{(\mu_1 - \mu_2)^2}$$

$$n = \frac{2(5.1)^2(1.645 + 0.842)^2}{(2)^2}$$

The hypothesis is one tailed because asthmatics are thought to have a decreased activity of G6PD compared with non-asthmatics. Hence the value of $z_{1-\alpha}$ is 1.645. If a two-tailed test were used $z_{1-\alpha}$ would be 1.960 (see below). From the formula above:

$$n = 80.44$$

Hence a sample of 81 subjects should be studied in each of the two groups. If the power were changed to 90% then the sample size increases to 111. If the difference is tested at the 1% level of significance and 80% power, the sample size is 102 in each group.

Two-tailed test

If we wanted to know if there was any difference in G6PD levels between asthmatics and non-asthmatics then the test would be two-tailed since no direction (asthmatics < non-asthmatics or non-asthmatics < asthmatics) was specified. The equation used would be identical to the one used above except that the α value would be for a two-tailed test, 95% $\equiv 1.960$.

$$n = \frac{2(5.1)^2(1.960 + 0.842)^2}{(2)^2} = 102.10$$

Thus a sample size of 103 in each group would be required for a two-tailed test as compared with 81 on a one-tailed test.

The design effect is 5.45. This implies that the variance of this sample is 5.45 times greater than the variance of a simple random sample of the same sample size. Suppose the standard deviation is 8.7, the variance is $(8.7)^2 = 75.69$. Multiplying 75.69 times 5.45 gives a variance of 412.51 for the cluster sample. The standard deviation is the square root of that number, 20.31.

The interpretation of this result is that given a median weight-for-height of 91.0, the 95% confidence interval lies somewhere between 91.0 ± 2(20.31) = 50.38 to 131.62. In conclusion, although the sample size was high, the estimates from the sample will not tell you very much due to the high standard deviation. These calculations were based on ROH of 0.05. In reality the homogeneity of these clusters could be much higher or lower.

Unfortunately, WHO does not discuss the problems of design effect. It is clear, as the above example demonstrates, that the large cluster size of 90 results in a very high variance. Assuming the same ROH of 0.05, deff would reduce to 3.45 with a cluster sample of 50, and to 1.95 with a sample of 20. If ROH was 0.01, the respective design effects would be 1.89, 1.49 and 1.19 for cluster samples of 90, 50 and 20.

The moral seems clear. If you are going to obtain the nutritional status or any other variable of a large number of subjects make sure you obtain your sample from a large number of clusters (villages).

How large a sample?

A study with an overlarge sample may be costly and unnecessary. On the other hand, a study with a sample that is too small will be unable to detect important effects.

The calculation of the sample size are based on 1. the level of significance (α), 2. the power of the test (β), and 3. for tests on means, the coefficients of variation of the changes and the percentage difference between group means (see Boxes 2.2 and 2.3 and Lemeshow *et al.*, 1990 for further details), or for proportions, the difference in proportions (see Box 2.4). The power of a significance test is a measure of how likely that test is to produce a statistically significant result for a population difference of a given magnitude (i.e. the power of a test is defined as the probability of correctly rejecting the null hypothesis, H_0, given H_0 is false).

The minimum acceptable level of significance used in scientific circles is 0.95. The test can be either one or two tailed (see Box 2.3). If for instance the hypothesis is that the treated will be better off than the placebo group, then the hypothesis is the alternative hypothesis (H_1) and the test is one tailed. The values chosen for the power of the test are commonly 0.80, 0.90 and 0.95. The problem of planning a study with low power means that the probability of finding no effect at all will be high even though an

BOX 2.4. Estimating the sample size for a proportion

A study is planned to estimate the prevalence level (proportion) of malnourished individuals for a predetermined (i.e. maximum) confidence interval.

The formula to calculate the confidence limit (CI) is:

$$CI = p \pm 2[p(100 - p)/n]^{1/2}$$

where CI is the upper and lower limits of the confidence interval, p is the percentage of malnourished individuals, $100 - p$ is the percentage of well-nourished individuals, 2 is the approximate value for the 95% confidence interval, and n is the number of individuals included in the sample.

The width W, of the confidence interval is expressed by the term $\pm 2\{p(100-p)/n\}^{1/2}$. Consequently $W = 2 \times 2\{p(100 - p)/n\}^{1/2}$. n can be derived by squaring both sides of the equation and multiplying by n:

$$W^2 \times n = 4 \times 4[p(100 - p)/n] \times n$$

$$n = 16p(100 - p)/W^2$$

To obtain n it is therefore necessary to choose a value of W and to know the value of p. Determining the value of p is, of course, the precise objective of the survey! However, it is not necessary to know the exact proportion and a rough guestimate (based on previous experience) will give a satisfactory result. For example, if the required precision for the estimate of change is $\pm 5\%$, width of confidence interval (W) is 10%, and estimated p value is between 30% and 35%:

$$n = 16 \times \frac{(35 \times 65)}{(10)^2} = 364$$

Using a value of p of 30% gives a sample size of 336.

If the malnutrition levels are 'patchy', then this heterogeneity will be reflected in the confidence interval. Cluster sampling will give a larger confidence interval than simple random sampling since in addition to the sampling error there will be variation between clusters which will contribute to the error. In the case of cluster sampling the formula used to calculate the confidence interval cannot be applied to estimate the required sample size. In practice, if a guesstimate of the prevalence of malnourished children as a whole in the population can be made the formula is:

$$n = 16cp(100 - p)/W^2$$

where c is a cluster factor. In most cases c is set to 2.

Using the same W and p values as before:

$$n = 16 \times 2 \times \frac{35 \times 65}{(10)^2} = 728$$

If 50 individuals are available to be measured in each village (cluster), then the total number of clusters is 15.

actual effect exists. Thus higher values of β and lower values of α will increase the sample size and therefore the cost of the study.

As indicated in Chapter 1, one should not be discouraged from attempting studies when a definitive answer cannot be found. Pilot studies are important for suggesting new avenues of research and for setting the assumptions for definitive studies of the types shown in the example provided in this chapter.

References
Aird, I. & Bentall, H. H. (1953). A relationship between cancer of the stomach and the ABO blood groups. *British Medical Journal*, i, 799–801.

Galton, F. (1889), *Natural Selection*. London: Macmillan.

Gore, S. M. & Altman, D. G. (1982). *Statistics in Practice*. London: British Medical Association.

Hennekens, C. H. & Buring, J. E. (1987). *Epidemiology in Medicine*. Boston: Little, Brown and Co.

Lemeshow, S., Hosmer, D. W., Klar, J. & Lwanga, S. K. (1990). *Adequacy of Sample Size of Health Studies*. Chichester: Wiley and Sons.

Sackett, D. L. (1979). Bias in analytic research. *Journal of Chronic Disease*, 32, 51–63.

Schor, S. & Karten, I. (1966). Statistical evaluation of medical journal manuscripts. *Journal of the American Medical Association*, 195, 1123–8.

Stephenson, L. S. (1987). *The Impact of Helminth Infections on Human Nutrition*. London: Taylor & Francis.

Stephenson, L. S., Latham, M. C., Kurz, K. M., Miller, D., Kinoti, S. N. & Oduori, M. L. (1985a). Urinary iron loss and physical fitness of Kenyan children with urinary schistosomiasis. *American Journal of Tropical Medicine and Hygiene*, 34, 322–30.

Stephenson, L. S., Latham, M. C., Kurz, K. M., Kinoti, S. N., Oduori, M. L. & Crompton, D. W. T. (1985b). Relationship of *Schistosoma haematobium*, hook-worm and malarial infections and metrifonate treatment to hemoglobin level in Kenyan school children. *American Journal of Tropical Medicine and Hygiene*, 34, 519–28.

3 *Biocultural studies of ethnic groups*

BARRY BOGIN

How does a biological anthropologist plan for, conduct, and analyze field research on people who belong to ethnic groups? The purpose of this chapter is to answer this question. The answer would be relatively straightforward, except that the term 'ethnic group' has no universally accepted definition within anthropology. A review of definitions used by anthropologists during the twentieth century includes the following selections (compiled by Crews and Bindon, 1991).

1. 'Ethnic groups are formed by virtue of community of language, religion, social institutions, etc., which have the power of uniting human beings of one or several species, races, or varieties' (Deniker, 1900, pp. 2–3).

2. 'The essential reality ... is not the hypothetical sub-species or races, but the mixed ethnic groups, which can never be genetically purified into their original components, or purged of the variability which they owe to past crossing. Most anthropological writings of the past, and many of the present, fail to take account of this fundamental fact' (Huxley and Haddon, 1936, p. 108).

3. 'When one uses the term "ethnic group", the question is immediately raised, "What does it mean? What does the user have in mind?" And this at once affords an opportunity to discuss the facts and explore the meaning and falsities enshrined in the word "race" and to explain the problems involved and the facts of the genetic situation as we know them' (Montagu, 1962, p. 927).

4. 'Ethnicity is a sociocultural construct that is often, if not always, coextensive with discernible features of a group of individuals. These features include, but need not be limited to, language, style of dress and adornment, religion, patterns of social interaction, and food habits' (Crews and Bindon, 1991, p. 42).

The first definition uses the concept of ethnicity as a social means to bridge biological differences between human beings. The notion of such differences at the level of *species* or race (i.e., sub-species) is explicit in this definition. The second definition rejects the idea that people belong

33

to distinct species, or even discernible races (i.e. sub-species). Rather, ethnic groups represent biologically heterogeneous mixtures. In this definition, ethnic groups are distinguishable due to their social context and not any biological distinctiveness. The third statement is not a definition. Rather, it is a plea to conduct the research needed to define the meaning of 'ethnic group', and to investigate the impact of ethnicity on human biology. That plea was made in 1962, and much research has been carried out since then. Based upon that research, the fourth definition represents the current 'state-of-the-art' use of the concept of ethnicity. In this usage, there is no biological contribution to ethnic identity. However, the 'discernible features' of each ethnic group may influence the biology of the members of that group. 'Language, style of dress and adornment, patterns of social interactions, and food habits' may influence many aspects of the biological structure of human groups. These include marriage and reproduction (i.e., genetic structure), nutrition, growth and development, body composition, transmission of disease, and mortality.

Despite the 90 years, and the apparent conceptual differences, separating these four definitions, there is a common theme that binds each to the others. That theme is the heart and soul of biological anthropology: the description and explanation of the physical differences between populations of people. This is both the *raison d'etre* and the oldest nemesis of the discipline. Biological anthropology exists to document and account for the variation within and between populations in physical growth, body composition, gene frequencies, physiology, demography, susceptibility to diseases, and other related characteristics. This is a legitimate area of research, and the results of such work have important implications for the medical, social, political, and economic welfare of all human beings. Yet, it is precisely because the scholarly work of biological anthropologists has practical social applications that the study of population, or ethnic, differences can also be used by some people to promote ethnocentrism, parochialism, intolerance for others, and discrimination against subordinate groups. At its worst, the study of ethnic differences stems from, and curries to, racism in its most pernicious intellectual form.

Historical background

Throughout the history of biological anthropology, most scholars equated ethnic groups with biological races. It was believed that people differ socially and behaviorally because of fundamental biological inequalities. Some people continue to write about ethnic groups and biological races as if they were synonymous. Scientifically, the production and maintenance of biological races requires an impediment to inter-breeding between human groups. Such groups, if they existed, could be properly called 'sub-species'. No sub-species of human beings exist, because there are no impediments to reproduction between any human groups. Thus, we could simply reject the concept of race and not discuss it further. However, to understand how biological anthropologists conduct research on ethnic groups today, it is necessary to discuss the history of the study of biological variation and the roots of 'scientific' racism. One could make the argument that much of the ethnic research done today is designed to counteract the racist history of the discipline. The following account is brief, often focused on the United States, and is meant to summarize a few concepts from an extensive literature on this topic (Gould, 1981; Brace, 1982).

Early historical accounts chronicle the discovery of groups of people who differed biologically and behaviorally from the explorers. Some of these accounts are imaginary. The Roman historian Pliny wrote about people of northern Africa, including the 'anthropophagi', cannibals who drank blood from human skulls, and the fantastical Blemmyae, whose heads were located below their shoulders (Friedman, 1981, pp. 9–12). Other accounts are more reasonable, such as ancient Egyptian sources that mention groups of very short stature people living near the head-waters of the Nile River, possibly the ancestors of central African 'pygmy' populations alive today (Hiernaux, 1974). The discovery of the pygmies posed many problems for Medieval European scholars. Peter of Croc asked in the year 1301 'Whether Pygmies be Men' (the title of one of his *quaestiones disputatae*). Peter decided that the pygmies were not human because of their short stature (biology) and their lack of morality (behavior). The only other question in dispute was whether the pygmies had fallen from grace or were a prehuman stage in the 'Great Chain of Being' (Friedman, 1981, p. 193).

Europeans also had to deal with the discovery of the 'new worlds', and their peoples, of Asia and the Americas. Such encounters continued into the twentieth century, with first contacts between Europeans and isolated groups of people living in highland New Guinea and in the rain forest of the Amazon. By now all such groups have been subjected to at least some degree of study by both untrained observers (e.g., the explorers, missionaries, and colonial officials) and professionally trained anthropologists.

Political and economic fortunes were to be made from the newly discovered worlds and peoples. Europeans, following the traditions of other imperialist peoples before them, won those fortunes by armed aggression, false promises, and the introduction of new infectious diseases (not always unintentional, but always devastating to the native peoples). Slavery and genocide became two of the by-products of European expansion. The justification for the physical and cultural carnage wrought upon the native peoples was always that *they* are different from *us*, and *they* are inferior to *us*. Europeans needed to create a category of the *other*, an inferior other, in order to subdue and dominate.

Others were easy to create because the people to be subjugated differed in some physical features, or some behavioral features, from the modal type of 'European' (of course Europeans differed among themselves in similar ways and this has been used as an excuse for warfare and exploitation within Europe as well). The creation of the *other* as an inferior being was aided by the Western philosophical notion of ideal types. This concept holds that there is a single perfect form of all creatures and objects, in this case the ideal type of human being. All living people are, to some extent, imperfect copies of that perfect form, Europeans held that they were closest to the ideal. Africans, Asians, Native Americans and other non-European peoples were, therefore, at a further distance in form and behavior from the ideal type. If one could measure, in precise 'scientific' terms, the amount of difference between the European form and the non-European form, then one would have an exact metric for determining the amount of inferiority of the native peoples. By the nineteenth century, some anatomists and anthropologists established methods for the systematic measurement of people and for their classification into discrete racial groupings. With the new methodology of anthropometry the era of scientific racism was born.

The inglorious history of racist science includes the origin and development of modern anthropology. This is not to impute all of anthropology as racist, but rather to remind students of anthropology of the parallel development of both racism and some anthropological methods that have been misused by racists. Another source of scientific racism was the incorrect interpretation of the new concepts of evolutionary biology. These misinterpretations contributed to a racist anthropology in that 'during the nineteenth century, there was general consensus in physical anthropology that the various "races" had been fixed entities [separate evolutionary lineages] for an immeasurably long period of time, possibly reflecting original creations – separate and unequal' (Brace and Hunt, 1990, p. 342). That belief came to its long overdue demise within anthropology (but not in other academic fields or in vulgar usage) with the publication of *Origin of Races* (Coon, 1962) and the storm of protest that followed (Livingstone, 1962; Brace, 1964; Mead *et al.*, 1968). It is now clear that whenever regional differences arose, *Homo sapiens* share a patristic affinity (Harrison, 1988, p. 324) and the biological variations of the species are shared to some extent by all modern groups of people (Lasker, 1984). In fairness, it may be more appropriate to use the term 'matristic affinity' to account for the biological unity of our species if the 'Eve hypothesis' is true. That hypothesis proposes that all living human beings are descended from a single female who lived in Africa about 200 000 years ago (Cann, 1988).

By the 1970s, advances in the field of population genetics demonstrated that human breeding populations showed more genetic variation within the group than between groups (Lewontin, 1974; Nei, 1975). Hartl notes that, 'genetic variation *within* [so called] races is so enormous that it all but swamps genetic differences *among* races' (1985, p. 370). This should have destroyed, for all time, the racist notions within biological and social science that differences between ethnic groups stem from separate evolutionary origins or that ethnic groups have evolved (since a common origin) uniquely different behavioral capacities. Unfortunately, racist ideas are still to be found in the recent literature. Psychologists Rushton and Bogaert (1989, p. 1216) claim that the greater rate of occurrence of acquired immune deficiency syndrome (AIDS) in people of African ancestry is due to their inherited biology. African peoples, as opposed to people of Asian and European ancestry who have lower rates for AIDS, are alleged to have genetically determined patterns of sexual behavior, intelligence (IQ scores), personality, and law abidingness that promote the spread of the disease. Most scholars simply dismiss these types of racist claims. However, fear of AIDS and ethnocentric views towards Africans may lead to a situation in which some non-Africans feel a sense of complacency that they are immune to the disease by virtue of genetic superiority.

We still live in an era where attempts are made to misuse science to support political and economic racism. This is hardly surprising given our history. In the early 1900s the United States Congress was debating and imposing restrictions on immigration of people from southern and eastern Europe. One reason for the restrictions was the alleged biological inferiority of these people. In 1912 and again in 1922, Boas published studies showing that the supposed 'racial' characters, believed to be biologically fixed markers within ethnic groups from Europe, changed or disappeared following migration to the US. Improvements in nutrition and new child rearing practices were some of the causes for these changes. In many cases, the newer southern and eastern European immigrants came to look more like the older immigrants from northern

Europe! Boas submitted an official report to the US Senate, but racist science, and the underlying racist fears that the new immigrants would take away some of the political and economic power of the older immigrants, won the day. The Immigration Restriction Act of 1924, targeted against southern and eastern Europeans, was passed into law. '"America must be kept American", proclaimed Calvin Coolidge as he signed the bill' (Gould, 1981, p. 232).

Immigration quotas are part of US policy today for peoples of Asian, African, and Latin American origin and, incredibly, some European ethnic groups such as Soviet Jews. The restrictions are based, officially, on economic and political considerations and past sources of immigrants, but to some extent these laws and regulations result from the persistence of belief that inferior biological characteristics are more common in some ethnic groups. A recent example is the false belief that casual social contact with Haitian immigrants to the US was a primary source of contagion to the AIDS virus. The public, and some politicians, demanded that Haitian immigration to the US cease. Since the Haitians in question are the decendants of African slaves, who were once deemed biologically inferior to Europeans, we see here an attempt to justify general racism against people of African origin by means of 'medical science'.

Definitions of ethnicity
It is against this historical background that the present discussion of methods for field research of ethnic groups begins. The first issue to be considered in field studies of ethnic groups is how to define ethnicity. The four definitions at the beginning of this chapter illustrate that in popular and anthropological usage, ethnic group, race, cultural group and other related terms are often used interchangeably. This reflects the fact that language is a social phenomenon and not an unambiguous reflection of reality. The study described in Box 3.1 illustrates this point in terms of popular classification of social races in Brazil.

Perhaps the most influential discussion within anthropology of the definition of ethnicity is found in Barth's book *Ethnic Groups and Boundaries* (1969). Barth and others in this edited volume criticize the traditional definitions of ethnic groups. These definitions attempt to identify such groups by what are believed to be their inherent characteristics; e.g., shared genetic history, shared cultural values and behavior, and shared language. Instead, it is argued that ethnic groups are not defined by any sort of natural diversity in biology or culture, 'but rather on the assignment of particular social meanings to a limited set of acts'

(Blom, 1969, p. 74). People create ethnic identities to suit the ecological, social, economic, and political conditions under which they live.

This view of the nature of ethnic classifications is supported by research showing that people can change ethnic groups within their own lifetime. The Fur and the Baggara, for example, are two ethnic groups in the western Sudan (Haaland, 1969). The Fur are farmers and the Baggara are nomadic cattle herders. They exchange complimentary goods, for example Fur millet for Baggara milk. The Fur farmers own some cattle, and Haaland found that when the number of cattle owned reaches five or more the Fur owner begins to give up the settled agricultural way of life. Fur farmers become Baggara nomads as the economic value of cattle herding exceeds that of farming. The Fur do not only change 'occupation', they also change social values, e.g., husband–wife economic roles, household organization, linguistic style, and attitudes toward ownership of land and material goods. Once the change is completed the people who may have been born and raised as Fur are Baggara.

Barth (1969) stated that the social construction of ethnicity is used to create boundaries between people so that the actors can 'categorize themselves and others for purposes of interaction' (pp. 13–14). 'The critical focus of investigation from this point of view becomes the ethnic *boundary* that defines the group, not the cultural stuff that it encloses' (p. 15). The interactions of actors on either side of the ethnic boundary are for biosocial purposes, including reproduction, production of basic foods and goods, religious behavior, political activities, and adaptation to environmental stresses (e.g., disease, heat or cold, malnutrition, etc.).

Ethnicity, by this definition, focuses on the management of social and biological behavior, behaviors that will have some effect on the people on either side of the ethnic boundary. For this reason, the study of ethnicity is still of legitimate concern for biological anthropologists. The field anthropologist must realize that the importance of the ethnic boundary is not that it isolates people into distinct groups (that is the outdated racist view of ethnicity), but rather that the boundary defines the types of interactions that will occur between people of different ethnicity who are all part of a larger social group.

An additional theme raised by Barth and his colleagues is that ethnic boundaries are often erected to define 'situations where social relations are characterized by inequality . . . between groups in society for political or economic access or dominance' (Nanda, 1991, p. 291). In one of the articles in Barth's book, Siverts (1969) describes the relations between the Maya and Ladino ethnic groups of southern Mexico in terms of this struggle for dominance. Siverts defines Mayans as those who speak a

40   *B. Bogin*

BOX 3.1. Classification of social races in Brazil (from Harris, 1970)

In the United States the classification of people into 'races' is based primarily on skin color and facial characteristics. These criteria divide people into two major groups 'whites' and 'blacks' (specific terms for people with Latin American or Asian facial characteristics are also used). In contrast, Brazilians use a much wider diversity of terms to classify people. Dr. Marvin Harris, then at Columbia University, conducted a field study to investigate the cognitive processes by which racial classification occurs in Brazil. He tested 100 native born Brazilians; men and women divided into upper, middle, and lower socioeconomic class and rural or urban residence from five states of the country. The test instrument was 'a deck of 72 full face drawings constructed out of the combination of three skin tones, three hair forms, two lip, two nose, and two sex types'. Ear shape, eye shape, amount of facial hair, facial expression, and all other features remained constant (Figure 3.1).

The subjects of the study gave a total of 492 different race names to the 72 drawings! The range of responses per person was 2 to 70 classifications. Harris notes that cross-culturally there is a positive correlation between the number of terms used and the precision of classification systems. This was not the case in Brazil. Disagreement between observers was the fundamental finding of this study. 'The most frequently employed terms were, in fact, applied to almost all of the drawings, and each of the drawings was identified by at least 20 lexical combinations.' Harris tried to find an order to the use of racial terms by counting the frequency of names used per drawing, but no clear pattern was discovered.

Brazil, like the US, has a large population of African origin (descendants of slaves) and of Native American Indians. Yet the two countries socially construct very different racial taxonomies. Harris speculates that the ambiguities of the Brazilian system may help to unify socioeconomic classes composed of people from diverse cultural backgrounds and with heterogeneous phenotypes. In the US the more rigid classification of race is used to fragment the lower social classes socially and to maintain their economic and political subordination to the upper classes.

Reference
Harris, M. (1970). Referential ambiguity in the calculus of Brazilian racial identity. *Southwestern Journal of Anthropology*, **26**, 1–14.

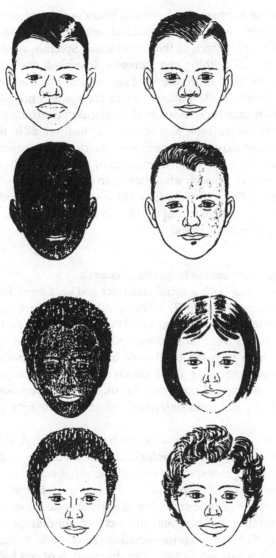

Figure 3.1. Examples of male and female drawings used in the field study of race names in Brazil.

Native American language and dress in traditional home-made Mayan clothing. Ladinos, who are the cultural descendants of the Spanish *conquistadores*, are defined as those who speak Spanish, wear manufactured Western-style clothing and 'pursue a Spanish-derived way of life' (p. 103). Mayans serve as producers of agricultural goods and consumers of industrial goods, while Ladinos act as traders and produces of the industrial goods and consumers of the agricultural products. Ladinos control most political, legal, and educational institutions in the region, and, thus, manage most of the power and wealth. By these means, which are a continuation of the colonial practices of the Spanish, they keep the Mayans subjugated. Mayans who attempt to change to Ladino ethnicity are derided by Ladinos as *revestidos* – 'those who have changed clothes' – and ostracized. Siverts finds that 'it takes at least two generations to get rid of the stigma of Indianhood' (p. 111).

Methods for conducting ethnic research

Viewing ethnicity as both a social construct and as 'a mask for confrontation' (Vincent, 1974, p. 377) allows the biological anthropologist to plan and execute field studies that avoid the pitfalls of past racist science. These older studies were mostly descriptive, and produced lists of trait inventories that were used to classify biological differences between people. Rather than using these differences to separate people into 'races', the clusterings of traits may be studied with methods suited to reveal the social and confrontational effects of ethnicity on human biology.

Living groups of people do show clusterings of biological characteristics. Groups of people differ biologically for two basic reasons. One is due to the random genotypic and phenotypic variability in non-adaptive traits that occurs in all populations. Accidents of history, including founder's effect and genetic drift in small isolated populations, migration, warfare, assortative mating, and all sorts of cultural practices may produce clusters of non-adaptive variation (Neel, 1970; Sokal and Uytterschaut, 1987; Gadjusek, 1990). The distribution of red hair color or surnames in the United Kingdom are specific examples (Sunderland, 1956; Lasker, 1985). Box 3.2 describes one such study.

The second reason that people differ is because they have made adaptations to local ecological conditions. These adaptations may be evolutionary, that is genetic, in nature due to the intensity and duration of natural selection for one or more adaptive traits. The distribution of malaria and the frequency of sickle cell alleles is the classic example of this type of adaptation (Livingstone, 1958). Or, the adaptations may be

BOX 3.2. Geography, language, and migration as causes of anthropometric variation in Kenyan ethnic groups (from Sokal and Winkler, 1987).

This article exemplifies one current statistical approach to the study of ethnic variation in biological characteristics, and shows that the causes of that variation are as much due to historical 'accidents' as due to any deterministic or biologically meaningful process. Dr. E-M. Winkler of the University of Vienna measured 19 head and body dimensions and five pigmentation variables on 4854 Kenyan children and adults belonging to 22 tribes and subtribes. With the assistance of Dr. R. R. Sokal of the State University of New York at Stoney Brook a sub-sample of these data, 958 people between the ages of 18 to 50 years from 15 tribes, was analyzed by spatial autocorrelation analysis. Spatial autocorrelation analysis 'attempts to describe the patterns of variation existing in the population sampled over a given area and to determine the factors that brought about the observed patterns'. The technique works by calculating 'the dependence of the values of a variable on the values of the same variable at specific physical distance'. For example, the autocorrelations for each of the 24 variables (e.g., head length or skin color) were computed for distances between tribes of 0–50 km, 50–125 km, 125–250 km, 250–500 km, and 500–1000 km. For the specifics of computation see the article.

The results show a significant spatial structure to the 15 tribes. Differences in all measurements reach a maximum by the 125–250 km distance, except for skin color which shows a cline from the northwest to the southeast of the area studied at all distances (i.e., maximum differences at 500–1000 km). Ten of the tribes are Nilotic speakers and five are Bantu speakers, and language and geography contribute significantly to biological variation among the tribes. This relationship is complex because both language and geography correlate significantly with variation in the 24 anthropometric variables, but there is no significant correlation between geography and language. The majority of the Nilotic and Bantu speakers tend to cluster in the same respective locales, but some notable exceptions such as the Maasai (Nilotic) and the Akikuyu (Bantu) settled a great distance from other tribes of their language group. The authors conclude that biological variation between these tribes is most likely due to 'considerable migration from different sources and in different directions'. Even the cline in skin color 'is probably due to the pattern of settlement of tribes with different shades of skin color rather than a selective cline'.

In addition to migration of tribes from place to place, the exchange of people between tribes, famines, and disease all contributed to the distribution of biological variation in these African populations.

Reference
Sokal, R. R. & Winkler, E-M. (1987). Spatial variation among Kenyan tribes and subtribes. *Human Biology*, **59**, 147–64.

BOX 3.3.   Ethnic and nutritional variation in hemoglobin S frequency in Liberia (from Jackson, 1990)

Like much of tropical west Africa, the nation of Liberia is holoendemic for falciparium malaria. This disease is the 'most consistent contributor to ill health in the indigenous pediatric and adult population'. Since Livingston's work in 1958 it has been known that the hemoglobin S (HbS) alleles exist at frequencies above 0.01 throughout the nation, but that there is also a northwest (NW) to southeast (SE) geographical cline in decreasing HbS frequencies (Figure 3.2). Dr. Fatimah Linda Jackson, of the University of Florida, hypothesized that the cause of this cline is due to ethnic differences in the amount and seasonal pattern of consumption of cassava, a root-crop staple in the diet. Cassava is the primary source of dietary cyanogenic glycosides. These are converted in the intestine to cyanide and are readily absorbed into the blood stream. In the blood, cyanide, thiocyanate, and cyanate may combine with proteins of the malaria parasite and inhibit its growth within the human red blood cell. In this way, cassava consumption may provide some resistance to malaria without the genetic load imposed by HbS alleles.

To test this hypothesis Jackson conducted a 15-month field study of 485 non-pregnant mothers. These women, from 17 indigenous ethnic groups

Figure 3.2. Map of Liberia with the geographical distributions of major ethnic groups, regions of low and high organic cyanogen consumption, and frequencies of hemoglobin S alleles.

from all regions of Liberia, were interviewed and observed on a daily, weekly, and monthly basis to document cassava consumption. They were tested for the presence of the HbS allele. The results, shown in Figure 3.2, support Jackson's hypothesis. In the NW region, low daily intake of cyanogenic glycosides is due to a seasonal consumption of cassava roots, mostly during the early rainy season and times of drought. In the SE and central regions, cassava is eaten year-round in large quantities by all age groups. Cassava is a staple weaning food for young children. Cyanogenic glycosides are also likely to be available in breast milk due to cassava consumption by nursing women. Furthermore, it is known that cyanide and thiocyanate can cross the placenta, so the fetus receives these compounds. Cassava-based foods are even given in small quantities ritually to newborns to affirm their membership in the Mano and Gio ethnic groups of the Central region.

Jackson's work shows clearly how ethnic cultural practices may have important consequences on the biology of those people.

Reference
Jackson, F. L. C. (1990). Two evolutionary models for the interactions of dietary organic cyanogens, hemoglobins, and falciparum malaria. *American Journal of Human Biology*, 2, 521–32.

ontogenetic in nature and due to environmental stressors that act during the growth and development of individuals within the same, or similar, population. Studies of children subjected to differing degrees of malnutrition, infectious disease, hypoxia, or psychosocial trauma illustrate this type of developmental adaptation (Bogin, 1988, pp. 105–59). Box 3.3 describes a recent field study of differences in diet and sickle cell allele frequency in several West African ethnic groups. This field study, like the majority of bioanthropological research, finds that most physically and socially important adaptations are ontogenetic in nature. In large part, these adaptations arise due to the ethnic boundaries that people erect. Those boundaries limit access to energy, material goods, or information required for existence. The degree of access influences the type of biological variation shown within each ethnic group (e.g., people with little access will often grow more slowly, be shorter as adults, and be less healthy at all ages than people with more access).

Biocultural methodology
When attempting to study the effect of ethnicity on either the adaptive or non-adaptive variations between human groups a *biocultural* approach is needed. The biocultural approach starts with the premise that all of human biology is the result of an interaction of genetic, developmental,

and environmental factors with cultural behavior. The examples of ethnic research in Boxes 3.1, 3.2, and 3.3 all used the biocultural approach. Each study, however, applied different methodologies to answer its research question. The biocultural approach allowed the investigators to ask questions and develop testable hypotheses that are more accurate reflections of reality, more sophisticated scientifically, and more satisfying intellectually, than older and simpler notions. Older notions, such as 'race', tried to separate artificially the biological and social aspects of people.

Field studies of ethnicity employing the biocultural approach require use of methods from both social anthropology and biological anthropology. Classic methods such as ethnography, participant-observation, anthropometry, and serology may each be used. Methods adopted from other disciplines, such as clinical interviews and examinations, physiological testing, nutritional assessment, biostatistics, and socioeconomic evaluation, may also be required. Methods based on new and emerging technologies may be needed, for example videotape recording of complex or fast-paced behavior and computer assisted analysis of multivariate and multidimensional ethnographic data. It is difficult to do justice to any or all of these methods in this brief chapter. It is possible, however, to get a 'feel' for how one may go about planning for a field study of ethnic effects by describing two cases. One is a successfully completed investigation of the effect of skin color, an ethnic boundary marker, on the incidence of hypertension in people from the United States and Brazil. The other case is a proposal for research on the influence of traditional Mayan medical and dietary practices on the growth and health of children in two Mayan villages in Guatemala.

Social class, skin color, and arterial blood pressure in two societies

This example, from the work of William Dressler (1991), illustrates the use of a sociocultural model to account for covariation in human skin color and blood pressure. The validity of the model is tested by collecting data on socioeconomic status and style-of-life and subjecting those data to a novel statistical analysis. The 'results are consistent with a model in which skin color and blood pressure are associated solely through a sociocultural process' (p. 60).

Dressler reviews literature that shows an association between the risk of increased blood pressure and dark skin color in populations from the United States and the West Indies. No clear genetic or biological factor, or factors, have been identified as causal for this association. Dressler addresses this association from a new social perspective. He argues that skin color is a *social variable* that serves as an ethnic boundary marker. In those societies using skin color to discriminate against people, darker skinned people are usually relegated to ethnic groups that are over-represented in the lower socioeconomic classes. Thus, skin color is an antecedent variable that influences education, income, access to health care, and other critical resources. But, this does not account for the association of skin color and blood pressure *per se*. To make this association Dressler draws on prior research that suggests that

> ... chronic blood pressure elevations may be related to a chronic struggle by the individual to acquire sufficient economic and social resources to achieve socially and culturally valued ends. These struggles are more or less difficult, depending on the ease of access to those resources. Access to those resources in some societies, including the United States, has been influenced by skin color. Therefore, darker skin color may be associated with higher blood pressure through this socio-culturally mediated pathway, rather than through a genetic mechanism (p. 62).

When the desire to achieve a certain level of economic and social well-being (i.e. a style-of-life) is constantly blocked, due to an ethnic boundary between people of darker and lighter skin color, there results a long-term stressful experience. One result may be higher blood pressure. Dressler developed a method to measure the amount stress people encounter due to their skin color and their social aspirations. The method groups people into three skin color classes: light, medium, and dark color as assessed by trained observers. These same people are interviewed to assess their style-of-life. The interview instrument includes items that examine the accumulation of material goods (TV, stereo, automobile, home owner-ship, etc.) and the adoption of behaviors that increase the consumption of information (travel, TV viewing, reading, etc.). People are also inter-viewed to examine their socioeconomic status (SES) and sense of per-ceived stress using standard inventories. Additionally, factors known to influence blood pressure, such as age, sex, body mass index (weight in kilograms divided by height in meters squared), calcium and fat intake, are also measured.

Dressler conducted two studies using this methodology. One was carried out in Brazil. There, 20 households each of seasonally employed sugar cane cutters, stably employed agricultural workers, factory workers, and bank employees were examined. The other study was done in a small southern city in the United States. In this case, two lower-income and two middle-income neighbourhoods were chosen and

Table 3.1. *Prevalence of hypertension in relation to life-style incongruity, defined by life-style and skin color, in Brazil and the United States*

| Life-style incongruity | Percentage hypertensive | | | |
| | Brazil | | United States | |
	n	%	n	%
Lighter skin color/lower life-style	48	18.8	43	6.8
Skin color = life-style	67	22.0	79	15.0
Darker skin color/higher life-style	13	46.1	59	23.7

From Dressler (1991).

a total of 181 households were chosen at random. With these data, Dressler was able to establish three categories of what he calls *life-style incongruity*. These include: 1. light skin color and lower life-style, 2. skin color equal to life-style, and 3. dark skin color and higher life-style. It was hypothesized that blood pressure would rise from category 1 to 3. Likewise, if the darker skinned person makes no attempt to participate in higher status groups (i.e., dark skin color and lower life-style), then there will be little incongruity or stress. In this case, there should be a lower incidence of hypertension.

Table 3.1 lists the results from the Brazilian and US studies. In both studies there is a statistically significant association between the percentage of hypertensive subjects and increased life-style incongruity. Covariates of hypertension, such as age, sex, body mass, and diet, do not eliminate the statistical significance of the life-style association. Dressler remarks that 'The combination of an ethnographic understanding of the social context of race or ethnicity in Brazil and the United States, along with the insights of social [conflict] theory, anticipate [these results]' (1991, p. 73). These studies are examples of the power of the biocultural approach, and related methodologies, to understand the nature of physiological variation between ethnic groups.

Traditional culture and biological adaptation in highland Mayan villages in Guatemala

The following is a proposal for research that was prepared by a cultural anthropologist, Dr. Barbara Tedlock, a student of medical anthropology, and by a biological anthropologist, Dr. Barry Bogin, a student of child growth and nutritional anthropometry. This combination of personnel was needed due to the biocultural nature of our problem. This proposal is presented here as a type of 'thought experiment' in the methods that are needed to settle a long-standing issue related to the Mayan people of Guatemala. That issue is the cause of the short stature of Mayan children and adults. It is often alleged that Mayan short stature is a genetic adaptation to centuries of poor nutrition following the Spanish *conquista* (Seckler, 1980). In this view, the Mayans are 'small but healthy'. Others argue that short stature is a symptom of malnutrition, not an adaptation. Mayans are suffering from malnutrition, and that suffering is due to social and economic boundaries that prevent Mayans from achieving the resources needed for healthy growth and development (Bogin *et al.*, 1989; Martorell, 1989).

To settle this issue, the proposed project will document some of the cultural processes that produce differences in nutritional status and growth in two villages in Guatemala. Both are Mayan villages in a general ethnic sense. (The definitions of Mayan and Ladino ethnicity discussed above for the research by Siverts (1969) may be applied here as well.) One village, however, is very traditional in terms of Mayan cultural ideology, social behavior, and practices toward food and health care. The other village is more acculturated to Ladino (Western) values and behaviors, and lacks traditional Mayan social structure and indigenous nutritional and health care practices. Our goal is to identify and document the effects of indigenous cultural knowledge and use of food on the nutrition and growth of Mayan children.

Previous research demonstrates that Mayan children, generally, suffer high rates of undernutrition, poor growth in height and weight, high rates of morbidity from infectious and chronic disease, and high rates of mortality (Bogin and MacVean, 1984; Bogin *et al.*, 1989). Even when compared with Ladino children from low SES families, Mayan children are, on average, more poorly nourished, smaller in size, and less healthy. Despite these generalities, some Mayan children are better nourished and grow larger than their peers, and many Ladino children, and some Mayan villages have lower average rates for childhood morbidity and mortality than others (Plattner, 1974; Bossert and Peralta, 1987). It is the goal of our research to determine, in part, why that is so. We hypothesize that traditional Mayan cultural knowledge and behavior ameliorates some of the negative effects of low SES on child growth and health.

Our research considers ethnicity in Guatemala in terms of the social construction and confrontation models of Barth and his colleagues. Since the Conquest, Mayans have lived through some four and a half centuries of ethnic subordination together with systematic political and economic disenfranchisement (Adams, 1970; Hawkins, 1984; Handy, 1984; Smith, 1985, 1988). This discrimination is still perpetuated by the politically dominant Ladino population (Manz, 1987; Davis, 1988). Yet, strong Mayan cultural identity has been maintained throughout this history (Carmack, 1988), in part due to the social boundary that both Mayans and Ladinos erect between their cultural identities. On the Mayan side, these social boundaries provided strategies for cultural survival, developed to deal with the new political and economic environment imposed by the conquerors. Our new research proposes that parallel strategies for biological adaptation and survival must also have evolved. As evidence of such biological strategies we may examine one standard measure of biological adaptation, population growth. Following initial declines after the conquest, Mayan population began to recover, indicating that biological adaptation to the new social environment of the Colonial period was taking place (Carmack *et al.*, 1982). Today, there is evidence for continuing increase in the size of the Mayan population (Arias, 1983; Bossert and Peralta, 1987).

Our work will be guided by the general hypothesis that two sociocultural factors are major determinants of biological adaptation (i.e., child growth and health). These factors are: 1. SES of the family, and 2. knowledge and use of appropriate health systems (traditional Mayan and/or Western biomedicine). These general factors are tested by the following three hypotheses:

1.   In both villages families of higher SES should have children with more favorable growth, better nutritional status, and lower incidence of morbidity than families of lower SES.

However, we suspect that there is a significant interaction between SES and knowledge and use of the appropriate health care systems. Thus:

2.   In the more traditional village we hypothesize that children will show higher levels of biological adaptation. This is due to the integrity of the indigenous Mayan health care system, and also access to Western biomedicine.

3.   In the more proletarian and acculturated community we hypothesize that the loss of knowledge and use of the indigenous health care system, and greater reliance on Western biomedicine, will result in less successful adaptation.

Our hypotheses do not predict that the indigenous health care system is superior to the Western biomedical system. Rather, if both systems are intact and functional there will be more alternatives for adequate diet and health care. This will be reflected in better growth. When only the Western biomedical system is available, Mayans, especially very low SES Mayans, will suffer due to financial costs and the conflict of values and behaviors between the Mayan clients and the Ladino health providers.

To test the specific hypotheses of our proposal, the research plan has eight aims, which are outlined as follows along with the methods to be used to accomplish these aims.

1. This is a two-year study in which we will establish ethnographic and physical anthropology fieldwork sites in two Mayan communities in highland Guatemala. One village is predominantly Mayan with a diversified local economy, traditional ideology and behavior, including the extensive use of pre-Hispanic calendars (solar, lunar, and 260-day), politically important land-holding patrilineages, and a large organized group of indigenous healers. The other village, though still Mayan in general character, lacks both traditional patrilineage structure and an organized group of healers and, due to its close proximity to Guatemala City (the capital of Guatemala), has become acculturated to Ladino values and is somewhat proletarian in character.

The method to be used to establish our fieldwork sites is to set up residence in each village, introduce ourselves to village officials (mayor, teachers, healers, physicians, etc.) and to village residents, and begin traditional ethnographic work collecting information through the use of participant-observation. The principle investigators (PIs) for this study have prior field experience in each of the villages and are acquainted with many of the residents, which facilitates this initial period of our research.

2. In each village we will identify as many families as possible with a child between six months and three years old. This age range is chosen because prior to age six months breast feeding can meet the nutritional needs of the infant. Breast feeding also provides the child with some immunity against infectious disease. After six months breast milk must be supplemented with other foods to meet growth demands. Reviews of the world-wide literature show that it is after age six months that growth failure due to malnutrition and infectious disease begins (Habicht *et al.*, 1974; Van Loon *et al.*, 1986). From this population of families, we will randomly select a sample of about 150 families. This sample will include children of both sexes, residents of village centers and rural hamlets, children of entrepreneurs, including traders and craftsmen, as well as of farmers and landless individuals, who are but marginally employed.

Practical and statistical considerations were used to choose the sample size of 150 families per village. Given the extensive nature of the data we wish to collect (see points 3–7 below) this is close to the maximum number of people we can visit, interview, and measure during a year. Our sampling design is to compare four cells; the two villages and a higher and lower SES group within each village. Using power analysis (Sokal and Rohlf, 1969; Hodges and Schell, 1988) we estimated that we would need a sample of 60 children per cell to ascertain statistically significant differences if cell means differed by one-half of a standard deviation. Therefore we require 240 children (4 cells × 60 per cell), but we will enroll 300 children due to the likelihood of drop-out and to increase the power of our statistical tests.

3. In both villages we will study the distribution of foods, including herbs (whether collected, grown, or purchased), food preparation and consumption within families, especially the intake of the study child. We will document both the everyday diet and the foods that are used seasonally and for special purposes, e.g., during illness, traditional festivals including celebrations of key days on the traditional calendar, Saints' Days, and marriages. Our purpose is to detail the diet of the study child and to determine when and why different categories of foods are prepared and eaten. We will photograph, label in Quiché and/or Kaqchiquel (the Mayan languages spoken in each village), and take appropriate plant samples for botanical identification of any unfamiliar foods or herbs for botanical identification.

One method employed for this aim is to follow individuals and record foods they collect or purchase and prepare for consumption by each study child, and then observe what each child actually eats. This is a time-intensive method and can only be accomplished for a small number of children, perhaps 10 children in each cell. For the entire sample we will inventory foods in the home, conduct 24-hour dietary recalls, and complete food frequency checklists (Quandt, 1987). These methods will give us a more global perspective of dietary patterns for our sample.

4. We will assess the biological adaptation of the child by taking anthropometric measurements and by observing changes in health status of the focus child.

Anthropometry will be used to measure both growth and nutritional status. Height, weight, arm and calf circumference, and skinfolds at the triceps, subscapular, iliac and medial calf sites will be measured. Under fieldwork conditions anthropometry is the most practical and reliable method for nutritional assessment (Habicht *et al.*, 1979). Height is an indicator of past nutrition and health history whereas weight relates to recent nutrition and health status. Circumference and skin folds are generally accepted indices of body composition, that is, lean body mass and fat mass. Lean body mass is a commonly used indicator for the body's reserves of protein and fat mass is an indicator of the body's reserves of energy.

Children will be measured twice during the year of fieldwork; once at the start of the study and the second time near the end of the study. Two measurements allow for the calculation of growth velocities which will give us a better understanding of the impact of periodic episodes of disease or undernutrition on the biological adaptation of each child. The data for achieved size and growth velocity for each study child will be compared with reference data for growth and nutritional status for Guatemalan populations of high and low SES. This comparison will permit us to evaluate the growth of each child in the context of the normal range of variation in Guatemala.

During the visit to each household to measure the study child, we will also assess recent morbidity of the study child by asking about episodes of illness experienced during the previous two weeks. For this we follow the methodology of Scrimshaw and Hurtado (1987; see Table 3.2 below), which was applied successfully by Frerichs *et al.* (1980) in rural Bolivia with Quechua-speaking people.

5. In both villages we will use informal conversations, in the appropriate Mayan language, together with questionnaires and structured interviews, in both Spanish and Mayan, to assess economic resources (ER) and social ranking of persons within the household (SR). ER includes the demographics of the family (size, ages, kinship relationships, etc.), occupations of the parents, income from all sources of work, rents, investments, etc., the value of all material assets (land, tools, machinery, business inventories, vehicles, animals, value of home) as well as physical characteristics of the home including type of construction, number of rooms, sanitary facilities, appliances, and books or other educational materials (information of this type was collected successfully by Annis (1987) in a Kaqchiquel village very similar in nature to the proposed study sites). SR includes level of formal education, literacy, participation in religious and social organizations, sponsorship of feasts, size and composition of kin and friendship networks, travel outside the community, and participation in Mayan versus Ladino institutions.

A person's, or a family's, ranking on ER and SR will be determined by the indigenous hierarchy of value ascribed to each variable in our questionnaires. Much of this indigenous valuation is known from previous research (Tedlock, 1982). The Guatemalan-specific categories will be expanded to include the standardized information on ER and SR

Table 3.2. *The major categories of information required for the stan-*
dardized assessment of SES, health status, nutrition, and patterns of
resort when seeking health care.

1. *Household composition:* number of peoples, ages, sexes, kinship relations.
2. *Type of housing:* single or multiple family use, number of rooms, construction
   material, sanitary facilities, water source, electricity, outdoor land use (garden,
   work areas, etc.)
3. *Socioeconomic status:* amount of education of each household member,
   employment of each household member, number of dependents, amount of land
   rented or owned, amount of land cultivated, amount of food stored in house,
   amount of food sold.
4. *Definitions of health and illness:* Ascertained by questions such as 'How does one
   (or, do you) know when a child is healthy?', 'How does one know when a child is
   ill?', 'What are the most common illnesses of children, and how are these
   caused?', 'What are the treatments and who provides treatment?'
5. *Foods eaten by healthy or sick children:* duration of breast feeding, weaning foods,
   foods for boys versus girls, qualities or functions of specific food items, food
   restrictions during illness, special foods for illness.
6. *Morbidity history of adults and children during prior 15 days:* the type of illness, its
   duration, perceived cause, treatment, and cost of treatment. Illnesses suffered
   during the fieldwork will be recorded as they occur for those households we follow
   on a daily basis. Preventative measures against disease will also be recorded.
7. *Inventory of household remedies:* presence of all remedies (both folk and
   'medical') used to prevent or cure illness, uses, sources, costs, and last usage (by
   whom, for what, etc.).
8. *Pattern of resort in use of health resources:* type of health specialist consulted
   (shaman, herbalist, public health worker, physician, etc.), order of consultations,
   reasons for consultations, unsolicited visits by health specialists to the home, use
   of official public health information and material (oral rehydration salts, family
   planning, vaccinations, etc.).

These categories conform to the data collection procedures recommended by the international research community (Scrimshaw and Hurtado, 1987). These data are collected by
open-ended interview and observation.

recommended by Scrimshaw and Hurtado (1987) in their protocols for
the assessment of nutrition and primary health care. Major topics covered
by the standardized assessment instrument are shown in Box 3.2.

Our measures of ER and SR will be used to construct a multidimensional index of socioeconomic status. The multidimensional index allows for
the construction of quantitative statistical models to test hypothesis. As
an example, the following model is offered. ER and SR may be used to
calculate two variables that Dressler (1982, 1988) refers to as 'accretion'
and 'discrepancy'. Accretion is the sum of ER and SR (i.e., ER + SR)
and is a measure of the total economic and social resources that an
individual or family has available. Discrepancy is the difference between

SR and ER (i.e., SR − ER) and is a measure of the incongruity between the social rank, or prestige, a person or family tries to maintain and the economic resources available to support that rank. We predict that both accretion and discrepancy have independent effects on biological adaptation. Higher values on accretion correlate positively with growth and health status and higher values on discrepancy correlate negatively with growth and health status. However, there may be interactions between these variables and the village of residence due to the traditional nature of one village and the acculturated nature of the other. Thus, a statistical model that considers accretion, discrepancy, and village of residence simultaneously should have the greatest predictive power.

These three variables can be analyzed efficiently by multiple regression. The regression formula is:

$$Y = a + b1(\text{accretion}) + b2(\text{discrepancy}) + b3 \text{ (village)}$$
$$+ b4(\text{accretion} \times \text{village}) + b5(\text{discrepancy} \times \text{village})$$
$$+ b6(\text{age}) + b7(\text{sex})$$

where Y is biological adaptation (e.g. growth status or morbidity frequency), and a is the intercept. Previous research in the acculturated village (Bogin and MacVean, 1987) shows that age and sex may be defined as covariates. This is a parsimonious and powerful model for assessing the effects of SES and village of residence on growth and nutritional status.

In a similar fashion other statistical models may be developed for variables constructed from any subset of the data base to test other hypotheses we are able to generate.

This index of SES, and its relationship with human biology, allows for a more 'fine grained' analysis than the more popular unidimensional model of SES, the Hollingshead index, which measures only a person's occupation and education. The multidimensional index of SES will allow us to place the child's state of biological adaptation in a wider social context of each family and village. Multidimensional models of SES have been applied successfuly in biocultural research on hypertension, depressive illness, growth, and health status in Jamaica, Mexico, Brazil, and the United States (Dressler, 1982, 1988; Dressler *et al.*, 1987a, 1987b; Malina *et al.*, 1985; Little *et al.*, 1988).

6. In both villages, when a study child becomes ill we will observe the utilization of traditional Mayan and Western biomedical health systems. We will research the process by which illness in the study child is diagnosed by his or her parents; what foods or medicines the parents use

to treat the illness; which types of health specialists are consulted, in what order; and what treatments are prescribed.

Cultural knowledge of health care will be gathered in three ways: 1. By in-home health interviews. Table 3.2 includes samples of the procedures used to collect this information. 2. By, early in our research, setting up meetings, in each community, with small focused groups of Mayan consultants during which they will be encouraged to talk freely, in either Mayan or Spanish, about health and nutritional issues. The participants will include rural health promoters, herbalists, midwives, bone setters, and women with children between six months and three years of age. These meetings will help us establish rapport with the community and our sample of families and assure the community of our sincere interest in traditional cultural behavior. 3. By observing what happens when a sample child becomes ill. The course of treatment will be documented, either with audio and/or video equipment (depending upon appropriateness and permission), and follow-up interviewing of the participants will be undertaken. Since there already exists relatively good documentation of many key Guatemalan herbs used as medicines and foods, our job will be to indicate when and how they are used by our study families. Upon occasion we expect to find unfamiliar herbs and will take plant samples for formal classification. By observing and taping healing sessions, and collecting illness narratives (stories set apart from the normal flow of conversation), we will be able to uncover cultural models of illness causation and knowledge specific to the inhabitants of each village concerning appropriate behavior when a child falls ill.

7. We hope to make audio and video tapes of health care interactions and practices, both in homes and during meetings and interviews with traditional healers and biomedical health practitioners. These tapes will serve as a data base for analysis of indigenous versus Western discourse concerning health and nutrition, as well as healer–patient interaction patterns, and health-seeking practices.

Our audio and video data base of treatments and conversations about health care and diet will help us to learn how Mayan individuals perceive and act in relation to health, nutrition, and illness. Subtleties in speech and behavior can be analyzed on tape that would be missed during the actual performance. After these tapes are analyzed, they will be archived so they may be used by other researchers both to check the results of this study and to conduct additional research.

8. The above information will be used to develop an interactive model of the system of biocultural adaptation to nutritional and disease stress in each village. The model will map the flow of significant cultural knowl-

edge and behavior relating to foods and health care that influence child growth, nutritional status, and morbidity.

Summary of the project

The major product of our research will be the demonstration of how ethnic boundaries produce variations in human biological adaptation. In this case the ethnic boundaries are the limits of nutritional and health care knowledge, availability, and use in a traditional Mayan community, Ladino culture in general, and a Mayan community that has acculturated to some Ladino values and behaviors.

The research goals, methods of fieldwork, and techniques for data analysis given here are meant to serve as a partial guide for students planning to undertake biocultural studies of ethnic groups. Many successfully completed field studies should also be consulted as one plans a new program of research (e.g. see past issues of the journals *Human Biology, The Annals of Human Biology,* and *The American Journal of Human Biology*). It is impossible to state if the proposed research proposal will be successful, for it has yet to be carried out. A proposal 'in progress' has been offered here to help the reader to think creatively about the type of ethnic research that needs to be done.

Conclusion

This chapter reviews briefly the earliest research on the biology of ethnic groups. That work was biased by ethnocentrism and the lack of knowledge or understanding of biological processes. The outcome of that early work was a xenophobic, or racist, interpretation of ethnicity and human variation. Some on-going research continues in this deplorable tradition. More recent research, which is biologically and culturally justifiable, focuses on the documentation and description of differences between populations. Little if any interpretation is offered as to the causes or meaning of biological variation, and no hypothesis testing is attempted.

An increasing tendency in current research focuses on the biocultural processes that produce the variations in adaptation between different ethnic groups. Such is the case for the work on skin color and hypertension and the proposed research on nutrition, health care, and child growth for Guatemala. The goal of current work is to conduct quasi-experimental studies. This is done by using the current distribution of social, economic, and environmental inequalities that are associated with ethnicity, to test hypotheses and predict the effects of ethnic social processes (e.g., the construction of and confrontations between ethnic groups) on human biology. It is hoped that some readers of this chapter will take up the challenge, and thereby contribute to an understanding of the role of ethnicity in biocultural variation and adaptation.

Acknowledgements
I thank Bernice Kaplan, Sandra Bogin, the editors, and anonymous reviewers for reading earlier versions of this chapter and offering many suggestions that improved the readability of the final text.

References

Adams, R. N. (1970). *Crucifixion by Power.* Austin: University of Texas Press.

Annis, S. (1987). *God and Production in a Guatemalan Town.* Austin: University of Texas Press.

Arias, J. (1983). *Demografica Guatemala: Una bibliografia anotada.* Primer suplemento. Guetamala: Universidad del Valle de Guatemala.

Barth, F. (1969). *Ethnic Groups and Boundaries.* Boston: Little, Brown.

Blom, J-P. (1969). Ethnic and cultural differentiation. In *Ethnic Groups and Boundaries*, ed. F. Barth, pp. 74–85. Boston: Little, Brown.

Boas, F. (1912). Changes in the bodily form of descendants of immigrants. *American Anthropologist*, **14**, 530–63.

Boas, F. (1922). Report on the anthropometric investigation of the population of the United States. *Journal of the American Statistical Association*, **18**, 181–209.

Bogin, B. (1988). *Patterns of Human Growth.* Cambridge: Cambridge University Press.

Bogin, B. & MacVean, R. B. (1984). Growth status of non-agrarian, semi-urban living Indians in Guatemala. *Human Biology*, **56**, 527–38.

Bogin, B. & MacVean, R. B. (1987). Growth status, age, and grade as predictors of school continuation for Guatemalan Indian children. *American Journal of Physical Anthropology*, **73**, 507–13.

Bogin, B., Sullivan, T., Hauspie, R. & MacVean, R. B. (1989). Longitudinal growth in height, weight, and bone age of Guatemala Ladino and Indian schoolchildren. *American Journal of Human Biology*, **1**, 103–13.

Bossert, T. J. & E. del Cid Peralta (1987). *Guatemala Health Sector Assessment 1987 Update.* Guatemala: USAID Mission in Guatemala.

Brace, C. L. (1964). The concept of race. *Current Anthropology*, **5**, 313–20.

Brace, C. L. (1982). The roots of the race concept in American physical anthropology. In *A History of American Physical Anthropology 1930–1980*, ed. F. Spencer, pp. 11–29. New York: Academic Press.

Brace, C. L. & Hunt, K. D. (1990). A non-racial craniofacial perspective on human variation: A(ustralia) to Z(uni). *American Journal of Physical Anthropology*, **82**, 341–60.

Cann, R. L. (1988). DNA and human origins. *Annual Review of Anthropology*, **17**, 127–43.

Carmack, R., ed. (1988). *Harvest of Violence: The Maya Indians and the Guatemalan Crisis.* Norman: University of Oklahoma Press.

Carmack, R., Early, J. & Lutz, C., eds. (1982). *The Historical Demography of Highland Guatemala.* Albany, NY: Institute for Mesoamerican Studies, SUNY-Albany.

Coon, C. S. (1962). *The Origin of Races.* New York: Alfred A. Knoff.

Crews, D. E. & Bindon, J. R. (1991). Ethnicity as a taxonomic tool in biomedical research. *Ethnicity & Disease,* 1, 42–9.

Davis, S. (1988). Introduction: Sowing the seeds of violence. In *Harvest of Violence: The Maya Indians and the Guatemalan Crisis,* ed. R. M. Carmack, pp. 3–36. Norman: University of Oklahoma Press.

Deniker, J. (1900). *The Races of Man.* London: Walter Scott.

Dressler, W. W. (1982). *Hypertension and Culture Change: Acculturation and Disease in the West Indies.* South Salem, NY: Redgrave.

Dressler, W. W. (1988). Social consistency and psychological distress. *Journal of Health and Social Behavior,* 29, 79–91.

Dressler, W. W. (1991). Social class, skin color, and arterial blood pressure in two societies. *Ethnicity & Disease,* 1, 60–77.

Dressler, W. W., Dos Santos, J. E., Gallaher, P. N. & Vitieri, F. E. (1987a). Arterial blood pressure and modernization in Brazil. *American Anthropologist,* 89, 398–409.

Dressler, W. W., Mata, A., Chavez, A. & Vitieri, F. E. (1987b). Arterial blood pressure and individual modernization in a Mexican community. *Social Science and Medicine,* 24, 679–87.

Freidman, J. B. (1981). *The Monstrous Races in Medieval Art and Thought.* Cambridge: Harvard University Press.

Frerichs, R. R., Becht, J. N. and Foxman, B. (1980). A household survey of health and illness in rural Bolivia. *Bulletin of the Pan American Health Organization,* 14, 343–55.

Gadjusek, D. C. (1990). Raymond Pearl Memorial Lecture, 1989: Cultural practices as determinants of clinical pathology and epidemiology of veneral infections; implications for predictions about the AIDS epidemic. *American Journal of Human Biology,* 2, 347–51.

Gould, S. J. (1981). *The Mismeasure of Man.* New York: Norton.

Haaland, G. (1969). Economic determinants in ethnic processes. In *Ethnic Groups and Boundaries,* ed. F. Barth, pp. 58–73. Boston: Little, Brown.

Habicht, J. P., Yarbrough, C., Martorell, R., Malina, R. M. & Klein, R. E. (1974). Height and weight standards for preschool children, How relevant are ethnic differences in growth potential? *The Lancet,* **April 6,** 611–14.

Habicht, J. P., Yarbrough, C. & Martorell, R. (1979). Anthropometric field methods: criteria for selection. In *Human Nutrition: A Comprehensive Treatise,* Vol 2, ed. R. B. Alfin-Slater & D. Kritchevsky, pp. 365–87. New York: Plenum.

Handy, J. (1984). *Gift of the Devil: A History of Guatemala.* Boston: South End Press.

Harrison, G. A. (1988). Part II: Human genetics and variation. In *Human Biology,* 3rd edn. ed., G. A. Harrison, J. M. Tanner, D. R. Pilbeam & P. T. Baker, pp. 147–336. Oxford: Oxford University Press.

Hartl, D. L. (1985). *Our Uncertain Heritage: Genetics and Human Diversity.* New York: Harper & Row.

Hawkins, J. (1984). *Inverse Images: The Meaning of Culture, Ethnicity, and Family in Postcolonial Guatemala.* Alburquerque: University of New Mexico Press.

Hiernaux, J. (1974). *The People of Africa.* London: Weidenfeld & Nicolson.

Hodges, D. C. & Schell, L. M. (1988). Power analysis in biological anthropology. *American Journal of Physical Anthropology*, **77**, 175–81.

Huxley, J. & Haddon, A. C. (1936). *We Europeans: A Survey of Racial Problems*. New York: Harper and Bros.

Lasker, G. W. (1984). The morphology of human populations. In *Estudios de Antropologia Biologica II: Coloquio de Antropologia Fisica Juan Comas, 1982*, ed. R. Ramos Galavan and R. Ma. Ramos Rodriquez, pp. 145–57. Mexico: Universidad Nacional Autonoma de Mexico.

Lasker, G. W. (1985). *Surnames and Genetic Structure*. Cambridge: Cambridge University Press.

Lewontin, R. C. (1974). *The Genetic Basis of Evolutionary Change*. New York: Columbia University Press.

Little, B. B., Buschang, P. H. & Malina, R. M. (1988). Socioeconomic variation in estimated growth velocity of schoolchildren from a rural, subsistence agricultural community in southern Mexico. *American Journal of Physical Anthropology*, **76**, 443–8.

Livingstone, F. B. (1958). Anthropological implications of sickle cell gene distribution in West Africa. *American Anthropologist*, **60**, 533–62.

Livingstone, F. B. (1962). On the non-existence of human races. *Current Anthropology*, **3**, 279.

Malina, R. M., Little, B. B., Bushang, P. H., DeMoss, J. & Selby, H. A. (1985). Socioeconomic variation in the growth status of children in a subsistence agricultural community. *American Journal of Physical Anthropology*, **68**, 385–91.

Manz, B. (1987). *Refugees of a Hidden War: The Aftermath of Counterinsurgency in Guatemala*. Albany: State University of New York Press.

Martorell, R. (1989). Body size, adaptation, and function. *Human Organisation*, **48**, 15–20.

Mead, M., Dobzhansky, T., Tobach, E. & Light, R. E., eds. (1968). *Science and the Concept of Race*. New York: Columbia University Press.

Monatagu, M. F. A. (1962). The concept of race. *American Anthropologist*, **64**, 919–28.

Nanda, S. (1991). *Cultural Anthropology*, 4th edn. Belmont, CA: Wadsworth.

Neel, J. V. (1970). Lessons from a primitive people. *Science*, **170**, 815–22.

Nei, M. (1975). *Molecular Population Genetics and Evolution*. New York, Elsevier.

Plattner, S. (1974). Wealth and growth among Mayan Indian peasants. *Human Ecology*, **2**, 75–87.

Quandt, S. A. (1987). Methods for determining dietary intake. In *Nutritional Anthropology*, ed. F. E. Johnston, pp. 67–84. New York: A. R. Liss.

Rushton, J. B. & Bogaert, A. F. (1989). Population differences in susceptibility to AIDS: an evolutionary analysis. *Social Science and Medicine*, **28**, 1211–20.

Scrimshaw, S. C. M. & Hurtado, E. (1987). *Rapid Assessment Procedures for Nutrition and Primary Health Care*. Los Angeles: UCLA Latin American Center Publications.

Seckler, D. (1980). Malnutrition: an intellectual odessy. *Western Journal of Agricultural Economics*, **5**, 219–27.

Siverts, H. (1969). Ethnic stability and boundary dynamics in Southern Mexico.

In *Ethnic Groups and Boundaries*, ed. F. Barth, pp. 101–16, Boston: Little, Brown.

Smith, C. A. (1985). Local history in global context: Social and economic transitions in western Guatemala. In *Micro and Macro Levels of Analysis in Anthropology: Issues in Theory and Research*, ed. B. R. Dewalt and P. J. Pelto, pp. 83–120. Boulder, CO: Westview Press.

Smith, C. A. (1988). Destruction of the material bases for Indian culture: Economic changes in Totonicapán. In *Harvest of Violence: The Maya Indians and the Guatemalan Crisis*, ed. R. M. Carmack, pp. 206–31. Norman: University of Oklahoma Press.

Sokal, R. R. & Rohlf, F. J. (1969). *Biometry*. San Francisco: Freeman.

Sokal, R. R. & Uytterschaut, H. (1987). Cranial variation in European populations: a spatial autocorrelation study at three time periods. *American Journal of Physical Anthropology*, **74**, 21–38.

Sunderland, E. (1956). Hair colour variation in the United Kingdom. *Annals of Human Genetics*, **20**, 312–33.

Tedlock, B. (1982). *Time and Highland Maya*. Albuquerque: University of New Mexico Press.

Van Loon, H., Saverys, V., Vuylsteke, J. P., Vlietinck, R. F. & Eeckels, R. (1986). Local versus universal growth standards: the effect of using NCHS as a universal reference. *Annals of Human Biology*, **13**, 347–57.

Vincent, J. (1974). The structuring of ethnicity. *Human Organization*, **33**, 375–9.

4 *Migration*

MICHAEL A. LITTLE AND PAUL W. LESLIE

Migration, movement, and human history

Migration can be considered a fundamental attribute of our species, since it has led to human habitation of nearly every terrestrial area in the biosphere. Throughout human history, people have moved their residences, often over vast distances. In paleolithic times, and up until the development of settled agricultural communities about 10 000 years ago, nomadic hunter/gatherers followed game and other resources in both the Old World and into the New World. This tendency for humans to move, either alone or in family groups, is great. The history of the migrations that led to the spread of humanity throughout the Old World (Africa, Asia, and Europe) during the paleolithic, and the later peopling of the New World (North and South America) and the Pacific, is complex indeed (Weiss, 1988). Evolutionary processes were imposed on migratory groups, who contributed genetic material to already resident populations and who exposed themselves to new environments with new selective pressures. Harrison (1984, p. 57) noted that: 'Movement is thus a first step in the diversification of human populations on the one hand, and the great homogenizer of populations on the other.'

McNeill (1978) proposed a four-fold pattern of migration that has characterized significant human movement since the rise of civilization and cities about 5000 years ago. Two kinds of movement (to cities and to frontiers) by two distinct groups (rural peasants and urban elite) constituted the four principal patterns of large-scale migration. Forty-five hundred years later, on the eve of 1492, this pattern was about to intensify. At that time, many continental populations had enjoyed relatively long periods of isolation that were about to end. In the 500 years since the discovery of the New World by Europeans, the redistribution of humanity through migration has been of dramatic proportions. Migration has transformed and continues to transform the spatial distributions of human populations on the planet.

Today, migration within national boundaries results from population pressures on existing land and other resources, as well as rising expectations associated with modernization and unequal distribution of wealth. Warfare and political conditions contribute to both national and international migration. During conditions of peace, most of the migratory streams within national boundaries are from rural to urban settings. International migration during peaceful times is most commonly motivated by the need for economic improvement, and people generally move from developing to developed nations.

What can be learned from migration studies?
Patterns of migration
Human migration takes many forms. At a relatively large scale, *colonization* or *frontier migration* refers to mass movements into uninhabited lands (Polynesian settlement of the Pacific islands) or inhabited lands of low population density (European settlement of North America). Some colonization may be forced, as in the involuntary transport of African slaves to the New World. At present, major colonization is occurring in the tropical forests of South America, where the combination of population pressure and economic opportunity has led to significant migration in Peru, Bolivia, Venezuela, and Brazil (Escobar and Beall, 1982; Lisansky, 1990). This pattern of migration involving 'the expansion of groups from their home ranges, if necessary replacing previous inhabitants of their new territory', has also been called *invasion* (Weiss, 1988, p. 130). Invasion usually leads to assimilation or destruction of the invaded population. Colonization of the New World by Europeans can be considered an example of an 'invasion'.

At the smallest scale, *matrimonial migration* (also called marital movement) involves one or both spouses changing residence in order to establish a household (Swedlund, 1984). Distances are usually small and spouses are most often drawn from the same ethnic communities. Matrimonial migration is one example of what can be called *local migration*; that is, where movement takes place over limited distances.

Between these extremes of scale are a variety of kinds of migration that involve movement over short or great distances. For example, *migrant workers* are usually unskilled, lower-socioeconomic-status invididuals who follow relatively low-wage employment opportunities. Migrant workers who cross international boundaries are often encouraged to migrate when low-wage labor is needed by the host country. Inevitably, many international migrant workers remain in the host country, while others return home. Other kinds of international migration frequently include members of *refugee populations* who are forced to leave their home country because of political or economic conditions. Palestinians in Jordan, Kurds in Turkey, and South Vietnamese in the US are examples of migrant refugee populations. International migration includes *illegal migrants* who cross national boundaries for the same reasons as do refugee populations (Clarke, 1984). Some individuals change residence after retirement. This form of *retirement migration* is quite common in the United States, where the 'sun-belt' states such as Florida, California, and Arizona have substantial populations of retired persons.

Involuntary or *forced migration* can take many forms: slavery, forced relocation for major construction, and sequestered prison and camp populations are a few examples. Migration is often used as a political tool to integrate peripheral areas into national systems. The settlement by Israel of the Gaza strip and the movement of ethnic Russians into Central Asia and the Baltic States are examples of this process. It is certainly not a new political device, since the Incas used this in their system of *mitimaes* or forced population movements to expand and control their Andean Empire in the fifteenth and sixteenth centuries (Steward and Faron, 1959, p. 127).

One of the most prevalent forms of migration occurring today is *rural-to-urban migration* (Bogin, 1988). Rural-to-urban migration takes place largely because individuals perceive that cities are centers of economic opportunity and excitement. The process, whether within or between national boundaries, has contributed to remarkable urban growth and widespread conditions of congested living that are unprecedented in human history. Migration from the countryside to the city dates back to the rise of cities in antiquity (McNeill, 1978), and has been one of the most common types of migration since that time. Indeed, until the last century, urban mortality rates were so high that most cities could not even maintain their sizes, much less grow, without substantial numbers of immigrants (Boyden, 1987, p. 164; McNeill, 1979).

Each of these kinds of migration leads to new environmental conditions (physical and social environments) that are experienced by a subset (migrants) of a given (home) population. The circumstances of migration can thus be identified as natural experiments to be exploited in a variety of ways. For example, what are the effects of environmental changes on migrant populations? Do migrants and those who remain at home differ in specific attributes? If so, how do they differ? What are the effects on the host population of an influx of migrants? These are just some of the broad questions that can be posed within a migration research design. More specific examples follow.

Early migration studies in biological anthropology

The earliest research design employing migrants was used by Franz Boas in the early years of the twentieth century and is described in Box 4.1. Boas' pioneering study of Eastern and Southern European migrants stimulated considerable later research on the physical characteristics of other migrant populations. For example, Harry Shapiro (1939) studied Japanese migrants to Hawaii, Marcus Goldstein (1943) measured Mexican-Americans, Gabriel Lasker (1946) studied immigrant and American-born Chinese, and Frederick Thieme (1957) investigated migration in Puerto Ricans. The research for each of these studies was based fundamentally on Boas' original design, where first and second generation immigrants were compared with the donor or sedente population to assess the effects of the new environment. Figure 4.1 is a diagram of the early Boas design that was followed by many other investigators. Most migrant studies have used this design in which a variety of migrant and non-migrants groups are compared. Modifications of the basic design include: comparison of ethnic groups with attempts to control for population genetic similarity or dissimilarity; comparison of individuals according to age at migration to examine developmental adaptation; comparison of sedentes (those who remain at home) with migrants to determine whether those who migrate are different than those who stay.

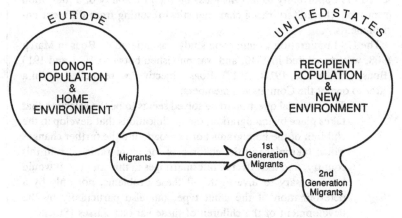

Figure 4.1. A schematic representation of Franz Boas' study of European migrants to the United States.

BOX 4.1.   Boas' study of European migrants to the United States

At the turn of the last century, the prevailing views on human populations were that races were fixed 'types', with adult forms strictly reflecting hereditary processes. Concepts of racial 'purity', Caucasian superiority, and the dangers of racial mixture were wisespread (Brace, 1982; Little, 1982). Few individuals challenged these popular ideas publicly; Franz Boas was one of these few. His interests in race, heredity, the environment, and social justice probably date back to his first work in anthropology in 1883–84 and even earlier to his youth in Germany.

The earliest research by Boas that revealed the effects of the environment on individual physical characteristics was his work on growth of school children in Worcester, Massachusetts. Begun in 1891, this was the first longitudinal study of child growth conducted in the United States (Tanner, 1981, p. 239). Although the study only continued for a year, Boas' careful measurements led him to observe considerable invidivual variation in what he called the 'tempo of growth' (growth rates). Some children grew more rapidly than others, and as Tanner (1981, p. 240) noted, Boas (1897) argued that social class differences in child growth resulted from variations in these growth tempos. Stocking (1974, p. 190) described Boas' perspectives clearly: 'Thus Boas did not conceive racial types in static terms, but as the products of the processes of heredity and growth within specific environmental situations.' Other studies by Boas of child and adolescent growth, including the establishment of United States growth standards, were published in the journal *Science* and elsewhere between 1892 and 1912. The ideas that came from all of this work were applied in the design of the study conducted for the US Immigration Commission. Here, Boas seized the opportunity to test his ideas on the probability of a 'deviation from type' (hereditary racial characteristics) resulting from different environments.

The US Immigration Commission study was intiated by Boas in March 1908, was completed in 1910, and was published between 1911 and 1913 (Boas, 1911, 1912a, 1912b, 1913). Boas' objectives were spelled out in a letter to one of the Commission members:

> The essential question to be solved seems to be the selection that takes place by immigration, the modifications that develop in the children of the immigrants born abroad, and the further changes which take place in the children of the immigrants born in this country; and the effect of intermarriages in this country. It would be necessary to investigate all these problems, not only by a determination of the adult type, but also particularly by the development of the children of these various classes (Stocking, 1974, p. 202).

Measurements of head dimensions and height were taken of thousands of Bohemians, Poles, Hungarians, Slovaks, Eastern European Jews, Sicil-

Photograph of Franz Boas taken in 1906, just two years before the European migration study was initiated (from Stocking, 1974).

ians, and Neopolitans, both immigrants in the U.S. and non-migrants abroad. Comparisons were made of children by age at immigration and those who were native US born. Boas' results laid to rest, forever, the belief that body characteristics were invariant (only under hereditary control), by demonstrating changes in head form (cephalic index), increases in height of immigrant children, and a negative relationship between family size and the average heights of children within the family (Boas, 1940).

References

Boas, F. (1987). The growth of children. *Science*, **5**, 570–3.

Boas, F. (1911). Changes in the Bodily Form of Descendants of Immigrants. Senate Document 208, 61st Congress. Washington, DC: Government Printing Office.

Boas, F. (1912a). Changes in the bodily form of descendants of immigrants. *American Anthropologist*, **14**, 530–62.

Boas, F. (1912b). *Changes in the Bodily Form of Descendants of Immigrants.* New York: Columbia University Press.

Boas, F. (1913). Veränderungen der Körperform der Nachkommen von Einwanderern in Amerika. *Zeitschrift für Ethnologie*, **45**(1).

Boas, F. (1940). *Race, Language and Culture.* New York: Macmillan.

Brace, C. L. (1982). The roots of the race concept in American physical anthropology. In *A History of American Physical Anthropology, 1930–1980*, ed. F. Spencer, pp. 11–29. New York: Academic Press.

Little, M. A. (1982). The development of ideas on human ecology and adaptation. In *A History of American Physical Anthropology, 1930–1980*, ed. F. Spencer, pp. 405–33. New York: Academic Press.

Stocking, G. W., Jr. (ed.) (1974). *The Shaping of American Anthropology, 1883–1911: A Franz Boas Reader.* New York: Basic Books.

Tanner, J. M. (1981). *A History of the Study of Human Growth.* Cambridge: Cambridge University Press.

Migration research problems and designs

Migration can be viewed as the movement of culture, disease, genes or people through space. Biological anthropologists have tended to concentrate on the study of the biobehavioral consequences of genetic migration (gene flow) and human migration of individuals, families, and populations, although there has also been considerable interest in migration as a vehicle for the transmission of disease (Kaplan, 1988). The methods of these major kinds of studies differ, necessarily, in their approaches and the questions that they ask.

Three principal kinds of investigation can be identified in biological anthropology. First, studies have been conducted to investigate *adaptation to environmental stress*. Boas' original study demonstrated developmental adaptation to new environments. Later investigations attempted to explain the reasons for these changes. Adaptation to high altitude hypoxia, to the heat and solar radiation of the tropics, and to the cold of the arctic have been studied often by use of migration designs (Lasker, 1969; Baker, 1977; Little and Baker, 1988). Second, migration has been used in studies of *health and epidemiology*. Infectious diseases, such as those that devastated New World populations after 1492, are often spread by migration, and new diseases based on modern life styles can be produced in migrants who become acculturated to these life styles (Kaplan, 1988; McNeill, 1978). Third, *genetic diversity* has been explored within a migration framework. Some time ago, Lasker (1960) outlined the importance of migration in studies of genetics and ongoing human evolution. As Roberts (1988, pp. 51–52) noted, migration transfers genes from one place to another, which contributes to genetic diversity of the host and migrant populations, and 'thus the vehicle for one of the major mechanisms of evolution, defined as change in gene frequency of the population'. Some studies have included the influence of gene flow on host and migrant populations, the founder effect in population variation, kin-based migration models, and other genetic models of migration (Dyke, 1984; Leslie, 1980; Mascie-Taylor and Lasker, 1987; Raspe, 1988; Termote, 1984). Discussion of each of these three kinds of investigation follows.

Adaptation to the environment

Adaptation is a fundamental concept in evolutionary theory, and is tied closely to related concepts of selection and fitness. *Relative fitness* is usually measured by reproductive success or performance, whereas *adaptation* is viewed as the way an organism is engineered to jibe with the environment. The states of fitness and adaptation are usually correlated. This results in *selection* of favorable (adaptive) characteristics which, if genetically based, are passed on to the next generation.

Adaptation need not be explored only in an evolutionary context. In order for biological evolutionary process to be demonstrated, it is necessary to show genetic change, which has proved extremely difficult to do even for systems in which it is quite clear that genetic mechanisms are in play. In a non-evolutionary sense, adaptation can be explored in the context of individual coping mechanisms, behavioral and biological adjustments to stress, relative merits of different strategies, physiological and behavioral flexibility, and developmental plasticity. Evolutionary process may be implied in these studies, but evolution is not always a major consideration. Scientific interests can be directed towards understanding mechanisms of adjustment or patterns of variation, including processes that contribute to or limit variation.

Baker (1976) suggested four research strategies or designs to investigate human adaptation to environmental stress. They included the following population–environment relations: 1. single population–single stress, 2. single population–multiple stress, 3. multiple population–single stress, and 4. multiple population–multiple stress. Each of the designs can be applied to migrant populations, but the first two designs can only be used with first generation migrants (single populations). The multiple population–single stress design requires all variables but one central stress to be controlled, whereas the multiple population–multiple stress design involves sorting out the interactions of a number of variable stresses. Examples of the four designs are given below.

Single population–single stress design

When this design is applied to migrants, it is necessary to initiate the study before or at the time of a move and then to follow the subjects longitudinally during the period of adjustment to the new environment. Few studies have used this design with migrants, since environmental change usually involves more than one stress, and it is difficult to be in a position to begin a study immediately before a move unless the move is planned. However, experimental designs of this sort have been applied where small numbers of subjects were moved temporarily to extreme environments. These included, among others, athletes at high altitude (Buskirk *et al.*, 1967) and military personnel in the hot desert (Baker, 1958). Population comparisons, of course, improve this, and the following, research design.

Single population–multiple stress design

The same conditions apply in this design as in the one above, and a planned move is generally required. The design is particularly useful for evaluating the effects of planned relocation of members of a population who are being moved because of some local danger or major construction. Scudder (1980) has described conditions of river-basin development and population relocation in Africa that would be especially appropriate for this design. One case that allowed application of this design was the relocation of Pacific Tokelauan Islanders to New Zealand after a hurricane hit the atoll in 1966 (Prior *et al.*, 1977). In this case subjects were tested and measured before they moved to New Zealand and followed in their new home.

Multiple population–single stress design

With several populations, comparisons can then be made that strengthen the design and the research results. When focussing on a single stress, it is necessary to control carefully other potential forms of stress. An example of this design, also cited by Baker (1976), is the work by Haas (1976) to assess the effects of high altitude on physical growth and psychomotor development of Peruvian infants (<2 years of age). An illustration of the design is given in Figure 4.2. The design controls for variation in rural–urban residence, socioeconomic status, and ethnicity, so that effects of the central variable, altitude, can be identified.

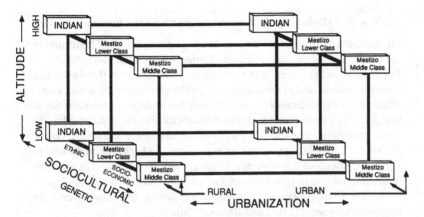

Figure 4.2. A research design of J. D. Haas that employs the comparative method and attempts to partition the effects of ethnicity, rural/urban residence, and altitude by use of migrants (from Baker, 1976).

Multiple population–multiple stress design

This research design is the most complex and thus requires careful control of the effects of multiple stress variables. However, the use of several migrant populations can be used to begin to sort out the importance of different stresses. Here, population comparisons can be used for their maximum power in defining patterns of adaptation to the multiple stresses of disease, limited nutritional intake, psychosocial disruption, and climate. An application of this design is described in Box 4.2. It is based on Geoffrey Harrison's (1966) strategy to test for altitude effects by comparing upward and downward migrants and their respective sedentary cohorts.

The more complex designs, especially the multiple population–multiple stress design, are more dependent on moderately sophisticated statistical analysis such as multiple regression and multivariate analysis of variance.

Health and epidemiology

Migration involves a variety of health threats. First, migrants can introduce infectious disease to members of their host or receiving population. This is particularly common during invasions, as in the movement of Europeans into the New World following its discovery by Columbus. Crosby (1972) has described the devastating transmission of smallpox (*viruelas*), measles (*sarampión*), and other epidemic diseases to Native

BOX 4.2.   High-altitude Andean migrants to the lowlands

The problem of how people are able to adapt to conditions of high altitude has been explored for many decades (Monge, 1948; Baker and Little, 1976; Schull and Rothhammer, 1990). Native populations from the Andes and Himalayas have lived at elevations over 3000 meters for thousands of years. What makes high altitudes of interest to human biologists is the presence of hypoxic stress from limited oxygen availability and the need to adapt to these conditions. For example, at 3000 meters above sea level, oxygen pressure is about 70 per cent of that at sea level, and culture can do little to buffer the effects of this reduction (Baker, 1976; Little, 1981). In sea-level residents, reduced oxygen pressure has a negative effect on most biological processes, from growth and the endocrine system to fecundity and the reproductive system (Little and Baker, 1976).

One useful design to study the effects of high-altitude hypoxia on both high-altitude and sea-level residents has been to compare migrants and sedentes (those who do not migrate). In contrast to the model of one-way migration, the high altitude model involves a two-way migration design. This design was first suggested by Harrison (1966) for the International Biological Programme high-altitude investigations which were conducted in Ethiopia, the Soviet Union, and the Andes. The design is a simple one as illustrated in the table.

	Low to high	High to low
Sedente	Low-altitude native	High-altitude native
Migrant	High-altitude migrant	Low-altitude migrant

Locating the migrant populations who have moved from sea level to high altitude or *vice versa* is often quite difficult, but comparison of these four groups provide a powerful tool for sorting out the effects of genetic and developmental adaptations. The design can be elaborated on by adding second generation migrants or by comparing different ethnic groups. A study of downward and upward migrants in Bolivia is described in the following study.

Haas (1980) wished to determine if prenatal growth was better in native Andeans than in Europeans at high altitude. He compared birth weights of infants born to mothers of European or native Andean (Quechua and Aymara) ancestry. European women who had migrated to high altitude (La Paz) as children or adults were compared with native Andean residents, and native Andean residents who had migrated to low altitude (Santa Cruz) as children or adults were compared with European women at low altitude. The author found that birth weights of native Andean infants were significantly greater than European infants at high altitude, but that both groups were roughly equivalent in birth weight at sea level. It was concluded that the native Andean mothers had a reproductive advantage over the European mothers, and that this adaptive advantage was based somewhat on heredity but also on a lifetime of exposure to high altitude.

Many other investigations have employed this migration design of upward and downward migrants as a means to understand better patterns of adaptation to environmental stress (Frisancho *et al.*, 1973; Beall *et al.*, 1977; Baker and Beall, 1982).

References

Baker, P. T. (1976). Evolution of a project: theory, method, and sampling. In *Man in the Andes: A Multidisciplinary Study of High-Altitude Quechua*, ed. P. T. Baker & M. A. Little, pp. 1–20. Stroudsburg, Pennsylvania: Dowden, Hutchinson & Ross.

Baker, P. T. & Beall, C. M. (1982). The biology and health of Andean migrants: a case study in south coastal Peru. *Mountain Research and Development*, **2**, 81–95.

Baker, P. T. & Little, M. A. (eds.) (1976). *Man in the Andes: A Multidisciplinary Study of High-Altitude Quechua*. Stroudsburg, Pennsylvania: Dowden, Hutchinson & Ross.

Beall, C. M., Baker, P. T., Baker, T. S. & Haas, J. D. (1977). The effect of high altitude on adolescent growth in southern Peruvian highlanders. *Human Biology*, **49**, 109–24.

Frisancho, A. R., Martinez, C., Velásquez, T., Sanchez, J. & Montoye, H. (1973). Influence of developmental adaptation on aerobic capacity at high altitude. *Journal of Applied Physiology*, **34**, 176–80.

Haas, J. D. (1980). Maternal adaptation and fetal growth at high altitude in Bolivia. In *Social and Biological Predictors of Nutritional Status, Physical Growth and Neurological Development*, ed. L. S. Greene & F. E. Johnston, pp. 257–90. New York: Academic Press.

Harrison, G. A. (1966). Human adaptability with reference to the IBP proposals for high altitude research. In *The Biology of Human Adaptability*, ed. P. T. Baker & J. S. Weiner, pp. 109–19. Oxford: Clarendon Press.

Little, M. A. (1981). Human populations of the Andes: the human science basis for research planning. *Mountain Research and Development*, **1**, 145–70.

Little, M. A. & Baker, P. T. (1976). Environmental adaptations and perspectives. In *Man in the Andes: A Multidisciplinary Study of High-Altitude Quechua*, ed. P. T. Baker & M. A. Little, pp. 405–28. Stroudsburg, Pennsylvania: Dowden, Hutchinson & Ross.

Monge, C. (1948). *Acclimatization in the Andes*. Baltimore: Johns Hopkins Press.

Schull, W. J. & Rothhammer, F. (eds.) (1990). *The Aymara: Strategies in Human Adaptation to a Rigorous Environment*. Dordrecht, The Netherlands: Klewer.

Americans at that time, and the reciprocal effects of syphilis transmission from the New World to the Old World. Second, if migrants move to cities, they are likely to move into ghettos that are the least desireable places, characterized by congested and unsanitary living conditions with high rates of epidemic infection. Flea-borne (plague) and louse-borne (typhus) epidemics in the cities of Europe are well documented in the historical literature (McNeill, 1976; Zinsser, 1935). Third, migrants move from a familiar area, to which they have some level of adaptation to the environment (including the infectious environment), to a less familiar area that will contain new diseases and conditions that may increase risks of new diseases. This is experienced particularly by pioneers who are moving from Brazilian cities to homestead the tropical Amazon forest (Moran, 1981). Fourth, if migrants move from underdeveloped to developed or non-Western to relatively prosperous Western conditions, they are likely to undergo dramatic changes in life styles that often have negative health consequences. Hence, despite the economic advantages gained by migration, there will be new health risks. These new health risks include the so-called 'diseases of civilization', such as diabetes (Weiss *et al.*, 1984; Zimmet *et al.*, 1977) and hypertension (Shaper *et al.*, 1969; Ward, 1983).

Two examples will be given of contemporary issues in health, epidemiology, and migration: the first is an infectious disease, acquired immunodeficiency syndrome (AIDS), and the second involves the health effects of life-style change and modernization.

AIDS

AIDS is a disease caused by a retrovirus called Human Immunodeficiency Virus Type 1 (HIV-1) that attacks the immune system. Its origins are obscure, but it probably arose in Africa in the 1960s through transmission, with modification, from *Cercopithecus* monkeys (Way, 1990). The rapid spread of AIDS around the world following its first appearance is a function of human mobility and migration. Mann and his colleagues (1988) identified three epidemic patterns of AIDS: 1. an epidemic among needle-sharing drug users and homosexual men (Americas, Western Europe, Scandinavia, New Zealand, and Australia), 2. an epidemic among heterosexual men and women (sub-Saharan Africa and the Caribbean), and 3. few cases with sporadic introduction (Eastern Europe, North Africa, the Middle East, Asia, and the Pacific). These patterns, especially (1) and (3), may be changing as the disease spreads. Patterns of direct transmission of AIDS include: heterosexual and homosexual intercourse, exposure to infected blood, and transmission from mothers to infants *in utero* or at the time of birth.

The threat of AIDS is truly great and global. With mortality rates at close to 100%, the ultimate mortality for the disease may equal the estimated prevalence and rates of HIV infection (Way, 1990). The Public Health Service estimate in 1988 of 1–1.5 million infected people in the United States (Berkelman and Curran, 1989) and the estimate of 2.5

million infected people in Africa (McGrath, 1990) identify this disease as one of the major pandemics of the century.

Although the immediate patterns of transmission are well known, the circumstances of past and present global spread of AIDS are still not clear. McGrath (1990) has argued for the need for a greater involvement in AIDS studies by biological anthropologists in areas of: 1. human biological variation and cofactors of infection and disease, 2. evolutionary impact, and 3. the interface between biology and behavior. These are certainly important areas of contribution by biological anthropologists; yet a fourth area of investigation is the exploration of patterns of infection transmission and spread at the national and international geographic scales. Such studies of migration and the spread of HIV in Africa (pattern 2), for example, would require information on human movement between rural and urban centers, and on sexual behavior of migrants, especially behavior linked to prostitutes in major population centers. In Asia and the Middle East, where pattern 3 exists (with limited cases and presumptive low infection rates) it is important to determine what are the routes of introduction from the West. It is important to determine, also, if the low rates are a function of limited migratory transmission of HIV, or if there is some degree of resistance among members of these populations that is linked to human biological variation and cofactors of infection and disease.

Research designs will require screening procedures and epidemiological tracking of HIV transmission, along with detailed studies of the biobehavioral and environmental circumstances of the infected people's lives. This is an area of research that will provide opportunities for the solution of both basic and applied scientific problems.

Health and modernization

Western life styles are associated with a syndrome of ill effects that include: obesity, affective and other eating disorders, diabetes, cardiovascular disease, hypertension, drug dependencies, anxiety, and other stress-related disorders (Kaplan, 1988). Such disorders appear for the first time in migrants who move from a native, non-Western environment to one that is characterized as modern or Western, and, particularly, an urban, Western environment.

Urban, Western life styles carry some risks to health and well-being and a number of sources of stress (Harrison and Gibson, 1976; Harrison and Jeffries, 1977). For example, inexpensive food tends to be readily available to middle or working class families, and excess intake is common. These diets tend to be high in calories, fat, salt, and protein, as well as additives and preservatives, some of which are believed to be toxic. The increasing reliance on 'fast food' enhances this trend. Nucleated settlement, clustered services, elevators, escalators, mass transit, and the automobile, have all contributed to reduce physical activity requirements in modern, urban residents. Reduced activity alone leads to poor physiological fitness, and when combined with excess dietary intake, contributes to poor cardiovascular health and frank disease. Another factor associated with urban Western residence is stress from the greater 'pace of life' that can lead to elevated catecholamines and hypertension (James et al., 1989). These generalized stresses are not well understood, and, in the case of migrant populations, may include components arising from subtle cultural differences. The effects of urban residence on disease patterns have been explored for decades as a part of epidemiological investigations. These disease patterns are complex and linked to socioeconomic and ethnic population distributions. Exposure to environmental pollutants, infectious disease vectors, and sociocultural conditions that are likely to predispose individuals to illness is more characteristic of city dwellers than of rural residents. However, even *rural* Western settings carry health risks that are not present in non-Western environments.

Box 4.3 provides an example of a series of migration studies of South Pacific islanders to New Zealand and Hawaii and the effects of modernization on members of these Polynesian populations.

Genetic diversity and microevolution

Evolution can be viewed as the process by which genetic variation within populations is converted to genetic variation among populations. Studies such as those described in Boxes 4.2 and 4.3 can tell us much about whether there are differences among populations that are likely to have arisen by genetic adaptation – that is, they deal with the *results* of microevolution. The study of migration and mobility is also important to an understanding of the *process* of microevolution. This entails identifying the factors that structure migration. That is, neither the destination of migrants nor the selection of migrants (who leaves and who stays) is determined by a random process in human populations. Because migration is nonrandom, its consequences are dependent on more than just the numbers of migrants or the rate of admixture.

The factors that structure migration include obvious influences such as geography (e.g., proximity of potential destinations) and political constraints on movement. There are also numerous, perhaps less obvious, social, economic, psychological, and cultural determinants of migration that may shape its microevolutionary consequences. Of particular interest to biological anthropologists are factors that result in migration that is not random with respect to biological attributes such as sex, susceptibility to specific diseases, fecundity, and consanguinity. Some individuals may be more likely to leave their natal population, or to remain in the receiving population, simply because they possess certain characteristics. This is termed *selective migration*. Macbeth (1984) considered possible causes of selectivity. Susanne (1984) discussed some of the problems in evaluating biological differences between migrants and sedentes. These include separating the effects of selective migration from those due to environmental differences and subsequent adaptation, and heterosis. The possibility of selective migration should be kept in mind when designing studies aimed at investigating the health consequences of migration, or adaptation to environmental stress, as discussed above.

Selective migration usually refers to migration that is nonrandom with respect to a specific characteristic of the individual, or to a set of associated attributes. Perhaps more generally important for genetic diversity within and among populations is migration that is influenced by genealogical relationships. This is *kin-structured migration*. When migration is nonrandom with respect to kinship (ancestry), even traits that have no bearing on the probability of migration may be nonrandomly distributed among migrants and non-migrants. This is the case because individuals tend to be genetically more similar to relatives, at all genetic loci, than they are to non-relatives. Groups of relatives are likely to be unrepresentative of the whole population with regard to large numbers of genetic traits. When such groups migrate, either all together or serially, the consequences of the movement for genetic diversity in both the sending and receiving populations are clearly different (usually greater) than if the same number of migrants were chosen at random. For example, movement of an extended family group might well completely remove a rare allele from a population. This important phenomenon is explored further in Box 4.4.

To a much greater extent than is the case with the studies of adaptation and health discussed above, the study of migration as part of the microevolutionary process entails the use of mathematical models. Of course, this in no way obviates the need for empirical studies. Rather, the formal models aid in the interpretation of relevant data; the data serve to test the models.

BOX 4.3.   Modernization and health in Polynesian migrants

One of the truly spectacular colonization migrations in the history of humanity was the settlement of the South Pacific islands by ancient Polynesian mariners (Bellwood, 1978; Baker, 1984). These migrations began about 3500 years ago, continued up to a few hundred years ago, and resulted in the settlement of every habitable island in the South Pacific. Pacific migrations continue, but now contemporary Polynesians are migrating out of traditional places to centers of Western society. They are undergoing what can be called a process of 'modernization'; that is, they are becoming acculturated by exchanging their traditional life styles for more Western life styles.

Migrant streams from Polynesian Islands have been directed toward many Western nations. Two important research projects focussed on migrants who moved from Tokelau Island to New Zealand (Prior *et al.*, 1977; Fleming and Prior, 1981) and from American Samoa to Hawaii (Baker *et al.*, 1987). In both cases, the migrants moved from a more central Pacific (and originally, Polynesian) site to a more distant Pacific (and Westernized) site. The research plans of the two independent projects were similar: to explore the effects of rapid environmental change and modernization, as brought about by migration, on health and disease in the new Westernized environment. The research designs were fundamentally the same as that illustrated in Figure 4.1, with some important variations.

Comparative analysis has always characterized research on migrants. However, several variants of this design were employed in these projects. First, the Tokelau Island Project was fortunate in having had access to data on Tokelau Islanders collected before a resettlement program was begun (a disastrous hurricane that struck Tokelau had made resettlement to New Zealand necessary). Hence, a *longitudinal design*, where individuals can be followed through time, was applied as a part of the research program. This allowed the investigators to observe directly the health effects on the migrants of continued residence in New Zealand. Second, the Tokelau Island Project was able to include other Polynesian populations with *varying degrees of acculturation* to Western society and technology. These included New Zealand Maori (most acculturated), Rarotongans (moderately acculturated), and Pukapukans (least acculturated). Third, degree of modernization or acculturation was considered also in the Samoan Project, but with a modified research design. Rather than using different Polynesian populations, comparisons were made of Samoans who had migrated to different places and who had achieved varying levels of assimilation and acculturation. Health and other attributes were compared along a gradient of acculturation that included Western Samoans (least acculturated), American (Eastern) Samoans, rural Hawaiian Samoans, urban Hawaiian Samoans, and Samoans living in California (most acculturated).

Modernization has brought to these Pacific migrants many of the so-called diseases of civilization: chronic obesity, hypertension, hyperuricemia, diabetes mellitus, and increased risk of cardiovascular disease. Two of these disorders (diabetes and chronic obesity) are in higher prevalence among Polynesians in the modern Western environment than they are among European-derived residents of the same environment. Such differences may have a hereditary basis that is linked to high fat intake and reduced physical activity. Longitudinal tracking of migrants and further comparison may well clarify some of the causes of poor health in Polynesian migrants.

References

Baker, P. T. (1984). Migrants, genetics, and the degenerative diseases of South Pacific islanders. In *Migration and Mobility: Biosocial Aspects of Human Movement*, ed. A. J. Boyce, pp. 209–39. Symposia of the Society for the Study of Human Biology, Volume 23. London: Taylor & Francis.

Baker, P. T., Hanna, J. M. & Baker, T. S. (eds.) (1987). *The Changing Samoans: Health and Behavior in Transition*. New York: Oxford University Press.

Bellwood, P. (1978). *The Polynesians: Prehistory of an Island People*. London: Thames and Hudson.

Fleming, C. & Prior, I. (eds.) (1981). *Migration, Adaptation and Health in the Pacific*. Wellington, New Zealand: Wellington Hospital Epidemiology Unit.

Prior, I. A. M., Hooper, A., Huntsman, J. W., Stanhope, J. M. and Salmond, C. E. (1977). The Tokelau Island migrant study. In *Population Structure and Human Variation*, ed. G. A. Harrison, pp. 165–86. Cambridge: Cambridge University Press.

Models

The classic migration models in evolutionary genetics were developed in the context of studies of genetic structure and subdivision of a species or regional population. These fall into two broad categories: isolation-by-distance (Wright, 1943; Kimura and Weiss, 1964; Malécot, 1969), and migration matrix models (Bodmer and Cavalli-Sforza, 1968, 1974). Cavalli-Sforza and Bodmer (1971) provided a good overview of the basic models and their variants. For discussions of the assumptions, advantages, and weaknesses of the models, see Harrison and Boyce (1972) and Jorde (1980).

These models have proven useful for exploring several fundamental aspects of the relationship between migration and microevolution: the rate of genetic change, the approach to equilibrium, and interaction with natural selection and random genetic drift. However, some of the models are very unrealistic for human populations, and all are largely descriptive. They say little about *why* movement takes a particular form, even when the fit between observation and a model's predictions are excellent.

BOX 4.4. Kin-structured migration

Following a period of growth, and typically in response to increasing tensions, Yanomama villages in the tropical forests of Venezuela and Brazil split, with one faction moving off to establish a new village (Neel, 1989). The factions consist of groups of relatives. During much of this and the past century, emigrants from the windward side of the small Caribbean island of St. Barthélemy went to one community on St. Thomas in the Virgin Islands; emigrants from the leeward side went to a different community on St. Thomas, socially isolated from the other but a scant three miles away (Benoist, 1964; Dyke, 1968). During the late nineteenth and early twentieth centuries, migrants from the Sicilian town of Cinisi concentrated in Midtown Manhattan; those from Sciacca, also in Sicily, settled in Norristown, Pennsylvania (MacDonald and MacDonald, 1964). The ancestry of Italian neighborhoods in dozens of other northeastern US cities can be traced to specific towns in southern Italy.

What these apparently disparate cases have in common is that migration is structured along lines of kinship. That is, genealogical relationships influence which members of a population will move together, which are more likely to emigrate at all, and where they are likely to go. This influence comes about because in many societies (perhaps most, historically) an individual's rights, obligations, expectations with respect to material resources, potential marriage partners, and more, are defined in large part by kinship. This both makes it necessary for people to remain in proximity to or in contact with their kin, and provides opportunities for migrants to get established in a new place. As explained in the text, kin-structured migration may have important genetic consequences.

Alan Fix has studied kin-structured migration in the Semai Senoi, who are swidden agriculturalists in central Malaysia. Communities of Semai periodically fission along kinship lines – a group of related migrants leave one settlement and join another. This pattern has been noted in other populations (e.g., several native South American societies). Fix was interested in the degree to which kin-structured migration might characterize pre-industrial populations and in the likely magnitude of the genetic effects of such migration. To investigate these questions, he used ethnohistorical, genealogical, and genetic data along with computer simulation.

Historical and genealogical data revealed the Semai migration pattern: its frequency, magnitude, and the way in which it is influenced by kinship and demographic constraints and other factors (Fix, 1974, 1975, 1981). Gene frequency data indicated that there was in fact substantial genetic differentiation among Semai settlements even though the settlements were not well isolated from one another (Fix and Lie-Injo, 1975). Computer simulation experiments are a means of exploring relationships that may be difficult to study by observation of real populations. Here, simulation led to a better understanding of the conditions under which kin-structured migration is likely to have important genetic consequences, and of the magnitude of those consequences. It also helped to clarify the ways in which kin-structured migration may interact with other factors (e.g. natural selection arising from epidemic diseases) (Fix, 1978, 1984).

Fix's work suggests that the potential importance of kin-structured migration is considerable. Since kinship has been so important in so many human societies, this sort of migration may have played an important role in human evolution. The arguments leading to this conclusion depended on integrating different approaches to the problem.

References

Benoist, J. (1964). Saint-Barthélemy: Physical anthropology of an isolate. *American Journal of Physical Anthropology*, **22**, 473–88.

Dyke, B. (1968). Numbers of potential mates in a small human population. Doctoral dissertation, Department of Human Genetics, University of Michigan, Ann Arbor.

Fix, A. G. (1974). Neighbourhood knowledge and marriage distance: the Semai case. *Annals of Human Genetics*, **37**, 327–32.

Fix, A. G. (1975). Fission–fusion and lineal effect: Aspects of the population structure of the Semai Senoi of Malaysia. *American Journal of Physical Anthropology*, **43**, 295–302.

Fix, A. G. (1978). The role of kin-structured migration in genetic microdifferentiation. *Annals of Human Genetics*, **41**, 329–39.

Fix, A. G. (1981). Endogamy in settlement populations of Semai Senoi: Potential mate pool analysis and simulation. *Social Biology*, **28**, 62–74.

Fix, A. G. (1984). Kin groups and trait groups: Population structure and epidemic disease selection. *American Journal of Physical Anthropology*, **65**, 201–12.

Fix, A. G. & Lie-Injo, L. E. (1975). Genetic microdifferentiation in the Semai Senoi of Malaysia. *American Journal of Physical Anthropology*, **43**, 47–56.

MacDonald, J. S. & MacDonald, L. D. (1964). Chain migration, ethnic neighborhood formation and social networks. *Millbank Memorial Fund Quarterly*, **42**, 82–97.

Neel, J. V. (1989). Human evolution and the founder-flush principle. In *Genetics, Speciation, and the Founder Principle*, ed. L. V. Giddings, K. Y. Kaneshiro & W. W. Anderson, pp. 299–313. New York: Oxford University Press.

There have been some attempts to develop more realistic models that include explanatory variables other than simple distance, or that are grounded in human behavior. One example, deriving from the isolation-by-distance approach, is the 'neighborhood knowledge' model for marital movement proposed by Boyce *et al.* (1967). It is based explicitly on behavior – more specifically, on the observations that people usually have a home base to which they return every day, and that, until recently, daily mobility was quite limited. Their model seems to be fairly successful in accounting for short range movement in sedentary agricultural populations (Swedlund, 1972). Another example, one that combines attributes of both categories of model, is that of Wood *et al.* (1985), who described a method for assessing the explanatory factors (sex, linguistic differences, population density, geographic distance, for example) that underlie an observed migration matrix. Nonrandom or selective migration has been modeled mathematically (Hiorns, 1985; Rogers and Jorde, 1987), and also has been subject to empirical investigation (see, for example, Lasker, 1952, 1954; Fix, 1978; Gibson *et al.*, 1984; MacBeth, 1984; Leslie, 1985).

Empirical approaches
In the absence of data pertaining to migration patterns, formal models are of limited utility. There are two general approaches to the empirical study of migration in relation to genetic diversity and evolution: 1. the direct approach of tracking actual migration events, and 2. indirect assessment by interpreting the prevalence of certain characteristics (e.g., allele frequencies) in the populations in question as reflecting migration in the past. Both approaches can provide information needed to evaluate the formal models and, with the help of appropriate models, to assess population affinities, rates of genetic exchange and differentiation, and other aspects of microevolution.

Places of birth and current residence are often included in birth, death, and marriage registers kept by civil and church authorities. Subject to the quality of the records, movement in and out of a population and interactions with other populations can be traced over long periods (see Chapter 7). Alan Swedlund and colleagues have made extensive use of historical records to investigate various aspects of the demography and population ecology of the Connecticut River Valley in Massachusetts during the eighteenth and nineteenth centuries. For example, Swedlund *et al.* (1984) studied community isolation and migration among a dozen townships over 60 years, in order to draw inferences about marital movement and genetic heterogeneity. To accomplish this, they used marriage records (which include birthplaces of spouses) in conjunction with Malécot's isolation-by-distance model. Parish records spanning more than 200 years and gene frequency data for recent generations provided the basis for analyzing the genetic structure and microevolution of the Åland Islands, Finland (Jorde *et al.*, 1982; Mielke, 1980).

Building on techniques of family reconstitution developed by historical demographers (e.g. Gautier and Henry, 1958), some anthropologists have been able to construct genealogies for entire populations and thereby to learn much about exchange between villages, islands, social strata, or other units that are relevant to population ecology and microevolution. The genealogical data add the important dimension of kinship to other information that can be derived from written records and from direct genetic and biometric assessment. Studies that have made use of extensive genealogies include those focussing on St. Barthélemy, French West Indies (Morrill and Brittain, in press; Leslie *et al.*, 1981), the Orkney Islands, Scotland (Brennan and Relethford, 1983; Brennan *et al.*, 1982; Boyce, 1984), Sottunga, Åland Islands (O'Brien *et al.*, 1989), the Utah Mormons (Jorde, 1989; Jorde *et al.*, 1983), and others.

An alternative to written records is retrospective data. Although data drawn from informants' memories are generally much more limited, written records are not always available, and information collected from oral histories and genealogies may allow reconstruction of past population movement. For example, it has been possible to demonstrate, for the Makiritare and Yanomama of South America, a good correspondence between village histories (movements and divisions) and genetic differentiation (Ward and Neel,1970; Neel *et al.*, 1977). The strength of the Tokelau project discussed in Box 4.3 was due in part to the researchers' ability to obtain genealogies with a depth of 10 to 15 generations, which link a large number of migrant and non-migrant Tokelauans (Ward *et al.*, 1980). These yielded essential information about genetic structure, which in turn provided an invaluable background for interpreting the epidemiological data (concerning blood pressure, diabetes, etc.) and for gauging the impact of the migrants' new environment.

In some cases, it is fruitful to combine a retrospective or ethnographic approach with information from written documents (Morrill and Dyke, 1980). This capitalizes on the strengths of each and permits some checks for consistency. Migration within and out of the genealogically related French populations of St. Barthélemy, French West Indies, and St. Thomas, US Virgin Islands, has been studied by this dual approach. The ethnographic data resolved some ambiguities in the written records, such as the coexistence of two people with the same name, or the problem of distinguishing individuals who actually emigrated from those who 'disappeared' from the records because they have remained unmarried and do not yet appear in the death register. The resulting genealogies could then be used to study a variety of problems associated with migration or marital movement (Dyke, 1971; Leslie *et al.*, 1981; Brittain, 1983).

The approaches discussed above involve determining specific migration events – that is, movement of known individuals or populations at certain times in the past. The observed migration patterns lead to inferences about genetic structure and microevolution. Conversely, at least to a limited degree, observed genetic structure (or proxies for it) contains information about patterns of interaction between populations (Boyce *et al.*, 1978). There is a large number of studies that seek to identify the origins of populations or the degree of past movement among populations by analyzing allele frequencies of those populations (see, for example, Devor *et al.*, 1984, and other papers in the same volume on the Black Caribs). Recent developments in molecular genetics now permit even more specific analysis based on nuclear and mitochondrial DNA.

To the extent that they are inherited in a regular manner, surnames can be used as convenient substitutes for alleles, and provide a basis for analyses similar to those that use gene frequencies. As originally made popular by Crow and Mange (1965), this approach focussed on the frequencies of *isonymous* (like-named) pairs. More recent generalizations to allow analysis of pairs of non-identical surnames have made the approach more powerful, but it remains dependent on assumptions that are often difficult to test. Perhaps the most important of these is that the surnames in a population are monophyletic – that is, a given name has only one origin. On the other hand, surnames are often recorded over many generations, while genotypes are not. Surname analysis is thus a useful tool for studying changes in movement and genetic structure through time. The single best overview of this method, including results of its application, is that of Lasker (1985). A symposium devoted to isonymy (Gottlieb, 1983) contains papers relevant to migration and microevolution.

A note on sampling

Because the determinants and consequences of migration are so complex, the circumstances associated with any given individual migrant are unique. Differences in the demographic, physiological and psychological statuses of individuals compound the variation arising from general differences in the social and physical environments of the donor and recipient populations. There are arrays of donor populations that send migrants to arrays of recipient populations. These migration events take place for individuals of all ages, through time, and the effects are felt across generations. Statistical control of such complex phenomena demands adequate sample sizes, and careful consideration should be given to sampling in any research design dealing with migration.

Random sampling is often identified as the ideal method. However, because of time and resource constraints, alternative strategies may be necessary. In some cases, greater analytic power may be obtained despite limited sample size by careful matching of migrants and sedentes. For example, studying pairs of like-sex siblings, one a migrant and the other not, may help control some of the myriad social, economic, and biological factors that can confound analysis. It is usually desirable to categorize individuals by age, sex, and perhaps other characteristics, but these variables should be chosen carefully, lest the subsamples become too small. Depending on the research problem, it may be wise to sample those age categories in which biological changes occur rapidly (infancy, adolescence) more intensively.

Future research in human biology and migration

The history of our species might be written as the spread and redistribution of populations over the Earth. Mobility, then, is an important attribute of humans. In many cases, it enables them to increase probabilities of survival and, more generally, to improve their states of well being. Human mobility does, however, entail risks, because it exposes individuals to new social and physical environments that test their ability to adapt to new circumstances. Hence, there are evolutionary implications in migration studies that provide a rich framework for a variety of investigations, and migration research will continue to contribute to our understanding of the limits to human adaptability. Cognizance of these limits is crucial in light of the global changes that all people now face.

International and internal migration during the twentieth century has been greater than at any other time in history. This trend is likely to continue for a variety of reasons. First, increase in human population numbers is exerting increasing pressure on land resources, leading to emigration. Second, population increases are contributing to depletion of resources (food, fuel, fiber, minerals, water) that reduce the ability of some areas to support large human populations. Third, these trends contribute to economic instability and an increasing disparity between economically advantaged and economically disadvantaged peoples, the 'haves' and the 'have-nots'. The outcome of such conditions is likely to be further increases in refugee migrants (who are often political refugees) and illegal migrants (who are often economic refugees) to Western nations, and greater rural-to-urban migration within Third World nations.

From the standpoint of biological anthropology, these observations have two implications. 1. Microevolution continues apace. Migration can be a potent force of evolution, and is now producing changes in the gene pools and genetic structures of many populations. And, new opportunities for natural selection arise as migration introduces old genotypes to new environments and as genes flow between newly neighboring sub-

86   *M. A. Little and P. W. Leslie*

populations. 2. Studies of the health and adaptability of refugee and other migrant populations may and should be an important applied research area in the future.

References

Baker, P. T. (1958). Racial differences in heat tolerance. *American Journal of Physical Anthropology*, **16**, 285–305.
Baker, P. T. (1976). Research strategies in population biology and environmental stress. In *The Measures of Man: Methodologies in Biological Anthropology*, ed. E. Giles & J. S. Friedlaender, pp. 230–59. Cambridge, MA: Peabody Museum Press.
Baker, P. T. (1977). Problems and strategies. In *Human Population Problems in the Biosphere: Some Research Strategies and Designs*, ed. P. T. Baker, pp. 11–32. MAB Technical Notes 3. Paris: Unesco.
Berkelman, R. L. & Curran, J. W. (1989). Epidemiology of HIV infection and AIDS. *Epidemiologic Reviews*, **11**, 222–8.
Bodmer, W. F. & Cavalli-Sforza, L. L. (1968). A migration matrix model for the study of random genetic drift. *Genetics*, **59**, 565–92.
Bodmer, W. F. & Cavalli-Sforza, L. L. (1974). The analysis of genetic variation using migration distances. In *Genetic Distance*, eds. J. F. Crow & C. Denniston, pp. 45–61. New York: Plenum.
Bogin, B. (1988). Rural-to-urban migration. In *Biological Aspects of Human Migration*, ed. C. G. N. Mascie-Taylor & G. W. Lasker, pp. 90–129. Cambridge: Cambridge University Press.
Boyce, A. J. (1984). Endogamy and exogamy in the Orkney Islands. *Northern Scotland*, **6**, 33–44.
Boyce, A. J., Harrison, G. A., Platt, C. M., Hornabrook, R. W., Serjeantson, S., Kirk, R. L. & Booth, P. B. (1978). Migration and genetic diversity in an island population: Karkar, Papua New Guinea. *Proceedings of the Royal Society of London*, B. **202**, 269–95.
Boyce, A. J., Küchemann, C. F. & Harrison, G. A. (1967). Neighborhood knowledge and distribution of marriage distances. *Annals of Human Genetics*, **30**, 335–8.
Boyden, S. (1987). *Western Civilization in Biological Perspective: Patterns in Biohistory*. Oxford: Clarendon Press.
Brennan, E. R., Leslie, P. W. & Dyke, B. (1982). Mate choice and genetic structure in Sanday, Orkney Islands, Scotland. *Human Biology*, **54**, 477–89.
Brennan, E. R. & Relethford, J. H. (1983). Mate choice and genetic structure in Sanday, Orkney Islands. *Annals of Human Biology*, **10**, 265–80.
Brittain, A. W. (1983). Migration from St. Barthélemy, French West Indies. PhD dissertation, Department of Anthropology, The Pennsylvania State University, University Park.
Buskirk, E. R., Kollias, J., Akers, R. F., Prokop, E. K. & Picón-Reátegui, E. (1967). Maximal performance at altitude and on return from altitude in conditioned runners. *Journal of Applied Physiology*, **23**, 259–66.
Cavalli-Sforza, L. L. & Bodmer, W. F. (1971). *The Genetics of Human Populations*. San Francisco: Freeman.

Clarke, J. I. (1984). Mobility, location, and society. In *Migration and Mobility: Biosocial Aspects of Human Movement*, ed. A. J. Boyce, pp. 355–70. Symposia of the Society for the Study of Human Biology, Volume 23. London: Taylor & Francis.

Crosby, A. W., Jr. (1972). *The Columbian Exchange: Biological and Cultural Consequences of 1492*. Westport, Connecticut: Greenwood Press.

Crow, J. F. & Mange, A. P. (1965). Measurement of inbreeding from the frequency of marriages between persons of the same surname. *Eugenics Quarterly*, **12**, 199–203.

Devor, E. J., Crawford, M. H. & Bach-Enciso, V. (1984). Genetic population structure of the Black Caribs and Creoles. In *Current Developments in Anthropological Genetics, Vol. 3. Black Caribs. A Case Study in Biocultural Adaptation*, ed. M. H. Crawford, pp. 365–80. New York: Plenum.

Dyke, B. (1971). Potential mates in a small human population. *Social Biology*, **18**, 28–39.

Dyke, B. (1984). Migration and the structure of small populations. In *Migration and Mobility: Biosocial Aspects of Human Movement*, ed. A. J. Boyce, pp. 69–81. Symposia of the Society for the Study of Human Biology, Volume 23. London: Taylor & Francis.

Escobar, M. G. & Beall, C. M. (1982). Contemporary patterns of migration in the central Andes. *Mountain Research and Development*, **2**, 63–80.

Fix, A. G. (1978). The role of kin-structured migration in genetic microdifferentiation. *Annals of Human Genetics*, **41**, 329–39.

Gautier, E. & Henry, L. (1958). *La Population de Crulai, Paroisse Normande*. Paris: Presses Universitaires de France.

Gibson, J. B., Harrison, G. A. & Hiorns, R. W. (1984). Migration and the structure of the Otmoor villages. *Annals of Human Biology*, **11**, 275–80.

Goldstein, M. S. (1943). *Demographic and Bodily Changes in Descendants of Mexican Immigrants*. Austin, Texas: Institute of Latin American Studies, University of Texas.

Gottlieb, K., symposium arranger (1983). Surnames as markers of inbreeding and migration. *Human Biology*, **55**, 209–408.

Haas, J. D. (1976). Prenatal infant growth and development. In *Man in the Andes: A Multidisciplinary Study of High-Altitude Quechua*, ed. P. T. Baker & M. A. Little, pp. 161–79. Stroudsburg, Pennsylvania: Dowden, Hutchinson & Ross.

Harrison, G. A. (1966). Human adaptability with reference to the IBP proposals for high altitude research. In *The Biology of Human Adaptability*, ed. P. T. Baker & J. S. Weiner, pp. 109–19. Oxford: Clarendon Press.

Harrison, G. A. (1984). Migration and population affinities. In *Migration and Mobility: Biosocial Aspects of Human Movement*, ed. A. J. Boyce, pp. 57–67. Symposia of the Society for the Study of Human Biology, Volume 23. London: Taylor & Francis.

Harrison, G. A. & Boyce, A. J. (1972). Migration, exchange, and the genetic structure of populations. In *The Structure of Human Populations*, ed. G. A. Harrison & A. J. Boyce, pp., 128–45. Oxford: Clarendon Press.

Harrison, G. A. & Gibson, J. B. (eds.) (1976). *Man in Urban Environments*. Oxford: Oxford University Press.

Harrison, G. A. & Jeffries, D. J. (1977). Human biology in urban environments: a review of research strategies. In *Human Population Problems in the Biosphere: Some Research Strategies and Designs*, ed. P. T. Baker, pp. 65–82. MAB Technical Notes 3. Paris: Unesco.

Hiorns, R. W. (1984). Selective migration and its genetic consequences. In *Migration and Mobility, Biosocial Aspects of Human Movement*, ed. A. J. Boyce, pp. 111–22. Symposium of the Society for the Study of Human Biology, Volume 23. London: Taylor & Francis.

James, G. D., Crews, D. E. & Pearson, J. (1989). Catecholamines and stress. In *Human Population Biology: A Transdisciplinary Science*, ed. M. A. Little & J. D. Haas, pp. 280–95. New York: Oxford University Press.

Jorde, L. B. (1980). The genetic structure of subdivided human populations: A review. In *Current Developments in Anthropological Genetics*, Vol. 1: Theory and Methods, ed. J. H. Mielke & M. H. Crawford, pp. 135–208. New York: Plenum.

Jorde, L. B. (1989). Inbreeding in the Utah Mormons: an evaluation of estimates based on pedigrees, isonymy, and migration matrices. *Annals of Human Genetics*, **53**, 339–55.

Jorde, L. B., Fineman, R. M. & Martin, R. A. (1983). Epidemiology and genetics of neural tube defects: an application of the Utah genealogical data base. *American Journal of Physical Anthropology*, **62**, 23–31.

Jorde, L. B., Workman, P. L. & Eriksson, A. W. (1982). Genetic microevolution in the Åland Islands, Finland. In *Current Developments in Anthropological Genetics*, Vol. 2, ed. M. H. Crawford & J. H. Mielke, pp. 333–65. New York: Plenum.

Kaplan, B. A. (1988). Migration and disease. In *Biological Aspects of Human Migration*, ed. C. G. N. Mascie-Taylor & G. W. Lasker, pp. 216–45. Cambridge: Cambridge University Press.

Kimura, M. & Weiss, G. H. (1964). The stepping-stone model of population structure and the decrease of genetic correlation with distance. *Genetics*, **49**, 561–76.

Lasker, G. W. (1946). Migration and physical differentiation. A comparison of immigrant and American-born Chinese. *American Journal of Physical Anthropology*, **4**, 273–300.

Lasker, G. W. (1952). Environmental growth factors and selective migration. *Human Biology*, **24**, 262–89.

Lasker, G. W. (1954). The question of physical selection of Mexican immigrants to the U.S.A. *Human Biology*, **26**, 52–8.

Lasker, G. W. (1960). Migration, isolation, and ongoing human evolution. In *The Processes of Ongoing Human Evolution*, ed. G. W. Lasker, pp. 80–8. Detroit: Wayne State University Press.

Lasker, G. W. (1969). Human biological adaptability. *Science*, **166**, 1480–6.

Lasker, G. W. (1985). *Surnames and Genetic Structure*. Cambridge: Cambridge University Press.

Leslie, P. W. (1980). Internal migration and genetic differentiation in St. Barthélemy, French West Indies. In *Genealogical Demography*, ed. B. Dyke & W. T. Morrill, pp. 167–77. New York: Academic Press.

Leslie, P. W. (1985). Potential mates analysis and the study of human population structure. *Yearbook of Physical Anthropology*, **28**, 53–78.

Leslie, P. W., Morrill, W. T. & Dyke, B. (1981). Genetic implications of mating structure in a Caribbean isolate. *American Journal of Human Genetics*, **33**, 90–104.

Lisansky, J. (1990). *Migrants to Amazonia: Spontaneous Colonization in the Brazilian Frontier*. Boulder, Colorado: Westview Press.

Little, M. A. & Baker, P. T. (1988). Migration and adaptation. In *Biological Aspects of Human Migration*, ed. C. G. N. Mascie-Taylor & G. W. Lasker, pp. 167–215. Cambridge: Cambridge University Press.

Macbeth, H. M. (1984). The study of biological selectivity in migrants. In *Migration and Mobility, Biosocial Aspects of Human Movement*, ed. A. J. Boyce, pp. 195–207. Symposium for the Society for the Study of Human Biology, Volume 23. London: Taylor & Francis.

Malécot, G. (1969). *The Mathematics of Heredity*. transl. D. M. Yermanos. San Francisco: Freeman.

Mann, J. M., Chin, J., Piot, P. & Quinn, T. (1988). The international epidemiology of AIDS. *Scientific American*, **259** (Oct.), 82–9.

Mascie-Taylor, C. G. N. & Lasker, G. W. (1987). Migration and changes in ABO and Rh blood group clines in Britain. *Human Biology*, **59**, 337–44.

McGrath, J. W. (1990). AIDS in Africa: a bioanthropological perspective. *American Journal of Human Biology*, **2**, 381–96.

McNeill, W. H. (1976). *Plagues and Peoples*. Garden City, New York: Anchor Books.

McNeill, W. H. (1978). Human migration: a historical review. In *Human Migration: Patterns and Policies*, ed. W. H. McNeill and R. S. Adams, pp. 3–19. Bloomington: Indiana University Press.

McNeill, W. H. (1979). Historical patterns of migration. *Current Anthropology*, **20**, 95–102.

Mielke, J. H. (1980). Demographic aspects of population structure in Åland. In *Population Structure and Genetic Disorders*, ed. A. W. Eriksson, H. R. Forsius, H. R. Nevanlinna, P. L. Workman & R. K. Norio, pp. 471–86. New York: Academic Press.

Moran, E. F. (1981). *Developing the Amazon*. Bloomington: Indiana University Press.

Morrill, W. T. & Brittain, A. W. (in press). *The Population of St. Barthélemy*. University Park: The Pennsylvania State University Press.

Morrill, W. T. & Dyke, B. (eds.) (1980). *Genealogical Demography*. New York: Academic Press.

Neel, J. V., Layrisse, M. & Salzano, F. M. (1977). Man in the tropics: the Yanomama Indians. In *Population Structure and Human Variation*, ed. G. A. Harrison, pp. 109–42. Cambridge: Cambridge University Press.

O'Brien, E., Jorde, L., Rönnlöf, B. Fellman, J. & Eriksson, A. (1989). Consanguinity avoidance and mate choice in Sottunga, Finland. *American Journal of Physical Anthropology*, **79**, 235–46.

Prior, I. A. M., Hooper, A., Huntsman, J. W., Stanhope, J. M. & Salmond, C. E. (1977). The Tokelau Island migrant study. In *Population Structure and Human Variation*, ed. G. A. Harrison, pp. 165–86. Cambridge: Cambridge University Press.

Raspe, P. D. (1988). Models of human migration: an inter-island example. In

Biological Aspects of Human Migration, ed. C. G. N. Mascie-Taylor & G. W. Lasker, pp. 70–89. Cambridge: Cambridge University Press.

Roberts, D. F. (1988). Migration in the recent past: societies with records. In *Biological Aspects of Human Migration*, ed. C. G. N. Mascie-Taylor & G. W. Lasker, pp. 41–69. Cambridge: Cambridge University Press.

Rogers, A. R. & Jorde, L. B. (1987). The effect of nonrandom migration on genetic differences between populations. *Annals of Human Genetics*, **51**, 169–76.

Scudder, T. (1980). River-basin development and local initiative in African savanna environments. In *Human Ecology in Savanna Environments*, ed. D. R. Harris, pp. 383–405. London: Academic Press.

Shaper, A. G., Leonard, P. J., Jones, K. W. & Jones, M. (1969). Environmental effects on the body build, blood pressure and blood chemistry of nomadic warriors serving in the Army in Kenya. *East African Medical Journal*, **46**, 282–9.

Shapiro, H. H. (1939). *Migration and Environment: A Study of the Physical Characteristics of the Japanese Immigrants to Hawaii and the Effects of Environment on Their Descendants*. London: Oxford University Press.

Steward, J. H. & Faron, L. C. (1959). *Native Peoples of South America*. New York: McGraw-Hill.

Susanne, C. (1984). Biological differences between migrants and non-migrants. In *Migration and Mobility. Biosocial Aspects of Human Movement*, ed. A. J. Boyce, pp. 179–93. Symposia of the Society for the Study of Human Biology, Volume 23. London: Taylor & Francis.

Swedlund, A. C. (1972). Observations on the concept of neighbourhood knowledge and the distribution of marriage distances. *Annals of Human Genetics*, **55**, 327–30.

Swedlund, A. C. (1984). Historical studies of mobility. In *Migration and Mobility: Biosocial Aspects of Human Movement*, ed. A. J. Boyce, pp. 1–18. Symposia of the Society for the Study of Human Biology, Volume 23. London: Taylor & Francis.

Swedlund, A. C., Jorde, L. B. & Mielke, J. H. (1984). Population structure in the Connecticut Valley, I: Marital migration. *American Journal of Physical Anthropology*, **65**, 61–70.

Termote, M. (1984). Migration as a selective factor in a population. Some considerations on the relation between migration and genetics. In *Population and Biology: Bridge Between Disciplines, Proceedings of a Conference*, ed. N. Keyfitz, pp. 133–46. Liège: Ordina Editions.

Thieme, F. P. (1957). A comparison of Puerto Rican migrants and sedentes. *Michigan Academy of Sciences, Arts and Letters*, **42**, 249–67.

Ward, R. H. (1983). Genetic and sociocultural components of high blood pressure. *American Journal of Physical Anthropology*, **62**, 91–105.

Ward, R. H. & Neel, J. V. (1970). Gene frequencies and microdifferentiation among the Makiritare Indians. IV. Comparison of a genetic network with ethnohistory and migration matrices; a new index of genetic isolation. *American Journal of Human Genetics*, **22**, 538–61.

Ward, R. H., Raspe, P., Ramirez, M., Kirk, R. & Prior, I. (1980). Genetic structure and epidemiology: The Tokelau study. In *Population Structure and*

Genetic Disorders, ed. A. W. Eriksson, H. R. Forsius, H. R. Nevanlinna, P. L. Workman & R. K. Norio, pp. 301–25. New York: Academic Press.

Way, A. B. (1990). Epidemiology and clinical picture of human immunodeficiency virus type 1 infection and the acquired immune deficiency syndrome. *American Journal of Human Biology*, **2**, 373–9.

Weiss, K. M. (1988). In search of times past: gene flow and invasion in the generation of human diversity. In *Biological Aspects of Human Migration*, ed. C. G. N. Mascie-Taylor & G. W. Lasker, pp. 130–66. Cambridge: Cambridge University Press.

Weiss, K. M., Ferrell, R. E. & Hanis, C. L. (1984). A New World Syndrome of metabolic diseases with a genetic and evolutionary basis. *Yearbook of Physical Anthropology*, **27**, 153–78.

Wood, J. W., Smouse, P. E. & Long, J. C. (1985). Sex-specific dispersal patterns in two human populations of highland New Guinea. *American Naturalist*, **125**, 747–68.

Wright, S. (1943). Isolation by distance. *Genetics*, **28**, 114–38.

Zimmet, P., Taft, P., Guinea, A., Guthrie, W. & Thoma, K. (1977). The high prevalence of diabetes mellitus on a central Pacific island. *Diabetologia*, **13**, 111–15.

Zinsser, H. (1935). *Rats, Lice and History: Being a Study in Biography, Which, After Twelve Preliminary Chapters Indispensable for the Preparation of the Lay Reader, Deals with the Life History of Typhus Fever.* Boston: Little, Brown.

5   Collection of human population genetic data

D. F. ROBERTS

There have been innumerable surveys of the genetic characteristics of human populations since the inherited basis of normal human variants was first established, and particularly with respect to serological features such as the blood groups, serum proteins, and red cell enzymes. For the ABO blood groups alone, the compilation by Mourant *et al.* (1958) included 3645 series (over one quarter containing more than a thousand subjects) published to the end of 1957. Indeed such enquiries in different peoples played an important role in discovering and exploring the extent of such polymorphisms, e.g. of the Hb A, S and C polymorphisms of the beta globin chain in Africans and the fast-moving variants of albumin in American Indians. Yet though texts of laboratory methods abound, setting out the technical procedures for typing the specimens, e.g. by Harris and Hopkinson (1976) for electrophoretic methods, Boorman and Dodd (1957) for blood grouping, and Davies (1986) for DNA analysis, there has been a singular paucity of guides to field survey methods in human genetics. This lack applies to surveys not only of normal genetic variants but of genetic diseases also.

For the former, Mourant (1958) discussed organisation for field research from the view point of the laboratory worker who received blood specimens, and who therefore was in a special position to appreciate the variety of ways in which blood samples, collected at the cost of very great effort, can be rendered unfit through mistakes made during collection or transport. He pointed out that simple tests can be done in the field, but recommended that specimens for fuller testing should be sent to a central laboratory. This remains true today for, though with modern transport and portable equipment much more sophisticated work can be done in the field than formerly, it is usually still more economical to bring the specimens to the laboratory, where a whole range of the necessary expertise, apparatus and reference specimens is concentrated and where the work on the specimens from one survey can be interdigitated with that on other material, than to transport several technical staff and equipment to a remote field location.

Human ingenuity, however, devises a range of solutions to the problem

92

of subjects in one place and sophisticated analytical procedures in another. For example, in a search for evidence of inherited heart conditions, a cardiological survey was made of the islanders of Tristan da Cunha; an electrocardiograph was sent to the island by sea, and when the subjects there were connected to it the readings were sent as radio signals to London where the ECG traces were recorded. The relative sparseness of the HLA data on remote populations is partly due to the difficulty in the earlier days of transporting material to the laboratory rapidly enough for testing. This difficulty was resolved by separating out the constituents of each blood specimen in the field, and refrigerating them separately, so that they arrived in the laboratory in a condition suitable for testing. But simple collection and despatch without such complex procedures can still provide a valuable contribution; e.g. DNA polymorphisms can be examined on blood specimens taken into EDTA.

Preliminaries

How the field worker obtains access to the required subjects and persuades them to participate is essentially an individual matter. It will vary according to their degree of sophistication and the country in which they live. As a preliminary, permission for the survey should be negotiated at the highest level, and here the good offices of contacts in, for example, the appropriate ministry of health or academy of science, possibly through international organisations (WHO, UNESCO), can be usefully sought. Members of faculty at a local university are often helpful in gaining official permission and in securing cooperation at the site of the study. Students from such an institution are sometimes a good source of intelligent field assistants. Sometimes the proposed survey can be integrated with another planned by one of these agencies or the appropriate ministry. But one encounters occasionally an individual in a position of responsibility who flatly refuses permission, and then some other channel has to be sought or the work deferred until he or she is replaced by another more sympathetic. Proof that such permission has been obtained is often required by grant-giving agencies.

Another essential preliminary is the detailed logistical discussion between the field worker and the recipient laboratory, so that agreement can be reached on the total number of specimens likely, the number to be included in each batch sent, the frequency of batches, and the maximal time that can be allowed between the collection of a specimen and its arrival in the laboratory, all of which are dependent on the capacities of the laboratory concerned. Agreement also needs to be reached on meeting the costs incurred by the laboratory. Since the work itself may be part of the laboratory's own research programme, it is often willing to make a contribution to these costs, but a busy laboratory receiving material from a large number of surveys at the same time may not have the resources to do so, and organisers of field surveys need to take account of these costs in applying for funding. Another point to be agreed is that of quality control, for every effort should be made to reduce errors. In obtaining blood specimens, a small percentage should be duplicated and assigned separate reference numbers and included in the main batches when sent, so that they provide a check on the technical consistency in the laboratory. But too many duplicates waste laboratory time and effort, and the number of duplicates that the laboratory will accept needs to be agreed initially. So too does the method of coding and labelling of specimens. It is important that laboratory testing should be done without knowledge of the identity of subjects, so avoiding any possibility of bias. The tests that the laboratory plans to carry out and that the field worker requires need to be agreed, for in some of these it may be necessary for the code to be broken either totally or partially before laboratory testing. One such case was the early HLA D locus testing using mixed lymphocyte culture methods which required family material and testing one specimen against another of known relationship. In the more common testing of X-linked characters the laboratory will require to know the sex.

Mourant (1958) gave valuable instructions on the collection, labelling, packing and transport of blood specimens. These were revised by him in 1964 and modified in the compilation by Weiner and Lourie (1969). These recommendations remain valid today, though with the increase in complexity of modern times, additional points need to be considered. Today, a number of grant giving agencies and ethical committees will only give finance or approval of the research if the subjects sign a form giving consent and stating that they understand the reason for their participation – quite impossible for many in a sophisticated society and a virtually insuperable obstacle when dealing with an illiterate people who believe more in magic than in science; a requirement where the pre-cautions of defensive medicine have in many cases exceeded the bounds of common sense. When dispatching the first batch of a survey to the recipient laboratory workers, the telegram or other message notifying them of this should state the airline manifest number, so that the recipients can alert the handling agents at the receiving airport and help them to watch for and trace the consignment. Some countries have strict controls over the import or export of blood specimens, and if such controls exist the procedure for meeting their requirements needs to be clarified before the field work starts, and possibly an expediting agent may need to be appointed at the point of shipment. Nothing is more frustrating than to see a consignment of blood specimens, which may have been collected at the cost of several thousand pounds, considerable discomfort to the field worker and inconvenience to the subjects, rotting on the floor of an airport office while officials who have not the slightest interest in it and who are ignorant of its scientific value argue about it.

Sampling

Mourant (1958) noted that the field worker who collected the specimens should choose his or her subjects 'according to certain well established genetical and statistical principles' but gave no guidance on these. Since for practical or financial reasons only rarely can a total population be investigated, it is necessary to take a sample. The method of sampling chosen is one of the most critical variables that can affect the results and their validity. Yet in almost no report of a serogenetic survey is the sampling procedure mentioned, and in the minority of reports where reference is made it is usually a simple statement to the effect that subjects were chosen at random, close relatives being avoided. Many a popular newspaper poll, and certainly most socioeconomic enquiries, have paid more attention to the sampling procedure than has the majority of serogenetic surveys. There is of course no one procedure that is to be recommended; it will depend on the questions that the enquiry is intended to answer.

The need is to draw inferences about a total population from data on only a part of it – the sample – and for this the sample must be representative of the whole population. To help define the population from which a sample is to be drawn, it may be useful to consult a national decennial census of population or other published sources if they exist, or obtain access to administrative records. Sometimes the units of such published sources are too large or otherwise unsuitable and for local studies some researchers have made their own census on the ground or estimated dwelling units from counts of roofs visible in aerial photographs.

A succession of samples would provide different estimates, variation among which (the sampling error) needs to be taken into account when interpreting the results and which needs to be reduced by one of the sampling techniques available. Examples of methods in common use are given in Box 5.1.

Sampling from human populations is different from sampling any other organism. An individual chosen for inclusion in the sample may be

BOX 5.1.   Some commonly used sampling methods

The simplest perhaps is to choose a set of members of the population at random, so giving a sample selected so that all members, and all samples of the same size from that population, have an equal chance of selection. This method allows valid inferences to be drawn about the population. In genetic studies of say the proportion of individuals in a population who have a certain blood group, the proportion in a random sample is an unbiased estimate of the population proportion, and the sampling error varies with the sample size.

Proportional sampling, with probability proportional to size of unit may be applicable in a situation where the population is stratified, for example by social class, or lives in villages, some large with many inhabitants, some small with only a few. In choosing villages or classes from which to obtain subjects, a member of one of the larger units would have a probability of selection different from a member of a smaller unit. In this situation, selection in two stages is obviously required, the first to establish the size of each unit to choose the sampling frame, the second to collect the specimens. Estimation of population parameters from such a survey involves weighting procedures.

Systematic sampling may be applicable, for example in new towns where households are arranged in a regular pattern, the households in each that are to be included and from which a member will be chosen are picked in terms of set distance and direction from each other starting from one chosen at random, and then one individual in each of those households gives a specimen. Here the population mean, but not the variation, can be estimated without bias.

Sample survey methods are given in standard texts such as Yates (1960), Stuart (1976) and Cochran (1977) and are dealt with in detail in Chapter 2.

References

Cochran, W. G. (1977). *Sampling Techniques*. 3rd edn. New York: Wiley.
Stuart, A. (1976). *Basic Ideas of Scientific Sampling*. 2nd edn. London: Griffin.
Yates, F. (1960). *Sampling Methods for Census and Survey*. 3rd edn. London: Griffin.

absent; he or she may be ill; he or she may simply refuse to give a specimen. Sampling frames and procedures for human studies must take this possible loss of subjects into account and allow for their replacement by methods that do not invalidate the sampling as a whole. On the other hand, offsetting these difficulties, human populations have many advantages for genetic surveys that no other species can provide. The subject can usually name his parents and his spouse and state where he was born, the number of his sibs and offspring. He or she can usually say if he/she is related to another subject and if so to what degree, and can often provide many other personal and family details that may be of relevance to the genetic data. Since all human populations have a social structure of some kind, he/she can usually say how he/she relates to it. In more sophisticated societies there may be records of births or family relationships. This material can often be used to good effect in sampling, as shown by the following examples, three of surveys of normal genetic variants, two with a clinical orientation. But in all cases the first requirement is to state the problem to be investigated and then choose the method that will provide the answer with the greatest accuracy.

Examples
i. Population studies
The Northern Nilotic survey

The study of the Northern Nilotics in the southern Sudan in 1953–4, carried out under extremely difficult field conditions, was among the first to attempt a coordinated modern approach incorporating genetics, physique, physiology, growth, health, demography and ecology to understanding the biological features of a population. A major object was to establish the extent to which variations in body form, and particularly of the extreme physique of the Nilotics, could be attributed to local variations in the way of life and other environmental factors. Comparison was therefore made of physique in three tribal samples of adult males who were living in identical habitats but with minor differences in their ways of life (Roberts and Bainbridge, 1963). First, some physiological control was needed to provide a quantitative validation of the apparent differences in their way of life, and this was done by means of a haematological survey, intertribal differences in haemoglobin level being found to correspond to those in their cattle wealth (Roberts and Smith, 1957). Secondly, to exclude the possibility that physique differences were due to genetic differences among the groups, some genetic control was needed, so a genetic survey was carried out (Roberts *et al.*, 1955). For this purpose it was important for the samples of the populations examined to be random but representative, that is to say with no preponderance of particular localities, families, or kindreds.

Northern Nilotic society is characterised by a lineage system in which every man traces his descent back to a lineage founder. The lineage system (Figure 5.1) regulates every detail of the daily life of the people, from the location of their huts in their villages, the grazing of their animals, the maintenance of law, to all the minutiae of daily interaction. Extremely important also, it regulates marriage in that a man may not

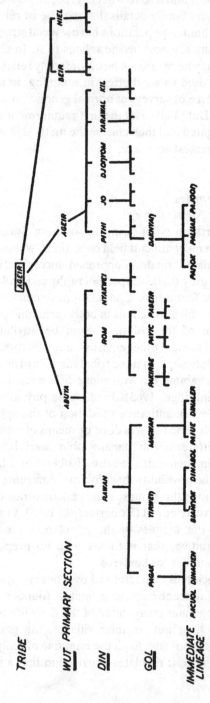

Figure 5.1. Choosing individuals for inclusion by making use of their immediate lineage ensures wide representation of the tribe and the avoidance of close relatives in the sample. Reproduced from Roberts (1956), by permission of Wayne State University Press.

marry a woman from the same immediate lineage, i.e. whose father is of the same lineage as himself. Not only does every man know his own lineage, but he also usually knows the lineages of all other men living in his own and adjacent villages, and knows members of his own lineage in distant villages. Here then the village and lineage system provided a useful structure by which a random sample could be selected to ensure that it consisted of unrelated individuals and was also representative of the population as a whole.

The Whickham survey

The genetic survey of Whickham carried out from July 1972 to June 1974 illustrates a procedure applicable in a more sophisticated population. The object was to assess by an epidemiologically oriented population survey the prevalence of hypothyroidism and other autoimmune thyroid disorders and ischaemic heart disease identify propositi for subsequent family investigations and, by including a wide range of serogenetic polymorphisms, to seek any gross associations that there were between the disorders and a variety of social and physiological variables on the one hand and normal genetic variants on the other (Tunbridge, 1976). The results of the gene frequency survey, though referred to in some comparative studies (Roberts, 1986), have not yet been published.

Whickham was chosen for its proximity to the base laboratories in Newcastle upon Tyne and for the characteristics of its population. Whickham Urban District, with at that time a total adult population of 20 680, occupies an area of approximately eight square miles to the southwest of Newcastle, and includes old mining communities, agricultural communities and residential suburban areas. The distribution of its population by age, sex and socioeconomic groups approximated that of Great Britain as a whole according to the 1966 national census figures, which were the latest available prior to the survey. Only the population over the age of 18 at the beginning of the investigation was considered for inclusion in the survey. The electoral register published in February 1972 was used as a sample frame. For the sample, the first subject was selected from the register at random and every sixth name was selected thereafter to give a one in six sample, with a size of approximately 3500. When a selected person had left the area or died but the family was still resident, no attempt was made to substitute for that person. When the selected person had moved away from his/her address and new occupants had moved in, one adult was chosen randomly from the adults in the new household. Having identified the selected person from the electoral register, the name of the general practitioner with whom the person was registered was obtained from the local medical council. The general practitioner then signed a standard letter to the individual inviting him or her to participate in the survey. When an individual was not registered with a general practitioner, the letter was signed by a clinical member of the survey team. The letter explained the purpose of the survey and invited the subject to attend a local health centre for interview. Persons who failed to reply or declined the invitations from their general practitioners were later visited by a member of the survey team to explain the purpose of the survey in more detail. No person was coerced into the survey and a refusal to participate after a visit from a member of the survey team was not pursued further.

Altogether a sample of 3538 persons was selected from the electoral role. One hundred and sixty-eight individuals were not available – 59 had died, 80 had left the area but their families were still living at the same address, 10 people were ineligible due to electoral role errors (mainly children), and 19 houses were empty. The available sample was thus 3370 persons, of whom 2284 readily accepted the initial invitation and were promptly seen in the survey. Members of the survey team personally made contact with all but 81 of the remaining 1086 selected individuals. Contact with these 81 failed despite at least four visits to their homes at varying times during the morning or evening of different days and these individuals were taken as refusals. Four hundred and ninety-five (46%) of the 1086 who did not respond to the initial invitation to participate in the survey did in fact agree to participate following the visit by a member of the survey team. The total number of persons seen in the survey was 2779 (82.4% of the available sample of 3370), and from all of them personal data, clinical histories and blood specimens were obtained and a wide range of physiological and serogenetic tests were carried out (Tunbridge, 1976).

The age and sex distributions of the sample are very similar to those of the population of Whickham urban district and Great Britain as a whole according to the 1971 census figures. Of the observed sample, 46.2% were male and 53.8% female, the mean age was 47.1 with a standard deviation of 16.4 years. Those who initially refused to take part were predominantly older females and those who initially failed to reply were predominantly young males. The occupation of the participant or her husband was used to group individuals by social class, the sample distribution of which closely resembles that of the total population of Whickham. The sample was essentially regional, for 29.1% of the participants were born in Whickham and 34% in the adjoining cities of Gateshead and Newcastle upon Tyne, while 81% of participants were born within the counties of Durham and Northumberland, as were 79% of their fathers and 81% of their mothers.

Thanks to this sampling procedure the sample is representative of Whickham not only demographically in its population composition but also in its genetic constitution. Its findings could be taken as reliable. They were of particular value in helping assess the biological meaning to be attached to the significant genetic associations reported in other studies, quite apart from contributing a number of useful suggestions requiring further study. For example, it could be clearly affirmed that in no serogenetic system was there any consistent gene frequency variation with age over the adult age groups covered. There were suggestions that social class I differed slightly in gene frequencies (for example with significantly elevated frequencies of the blood group phenotypes P1−, A_1, and rhesus dd, though the results were not equally significant in the two sexes); that there was differential fertility in Lutheran and rhesus D (shown by the number of sibs after correction for the age of the propositus), and in PGM (in the number of children after similar correction); that weight and weight for height corrected for age was slightly associated with the Kidd phenotypes.

The Basque survey

A third example is of a survey undertaken to establish the gene frequencies of a given ethnic group with a view to establishing affinities with and differences from other populations. With such an object particular attention is required in the selection of the subjects. Populations today are characterised by migrations of individuals, often temporary, often permanent, and the presence of immigrants in a sample will distort any estimate of the genetic constitution of the population under examination. Any appraisal of population affinities must take into account the dynamic nature of the genetic constitution.

According to the Spanish system of surnames, in which each individual carries the name of his father and mother, most people know the eight names of their four grandparents. This system can be used in sampling. In a survey of the HLA gene frequencies of Spanish Basques by Calderon *et al.* (1993), subjects born in the four Spanish Basque provinces were included if all eight grandparental surnames were Basque, or if seven were Basque and the eighth was from an adjacent province. The frequencies of HLA antigens A3, A28 and A29 differed significantly from those in the Basque sample of de Masdevall *et al.* (1982), and a multivariate analysis showed the latter sample to be situated midway between the former sample of Spanish Basque and non-Basque data. de Masdevall's sample consisted of residents of Bilbao with four Basque names (i.e. both names of both parents) and with three generations of Basque antecedents. With the difference between the two sampling methods, it is difficult to decide whether the difference in results between the two samples represents their different areas of residence (one urban and one from the more rural provinces) or the possibility that the urban sample may include a greater proportion of immigrant, non-Basque genes than the other.

ii. Family studies

A different approach to data collection is required when a rare genetic disorder is being investigated in a given population to establish its mode of inheritance, or when the object is to examine the extent of the genetic contribution to the etiology of the disorder. These studies require the collection of family data, and these present particular problems. The pattern of occurrence of affected individuals in pedigrees provides a useful guide, and critical information comes from comparison of observed ratios in sibships with those expected on Mendelian hypotheses. But because human families are small, the theorectical proportions of the genotypes that segregate in sibships rarely meet the proportions expected on a given mode of inheritance, so data from many families have to be combined. In dealing with a recessive disorder, the fact that the parents are heterozygous carriers is only established by the observation of an affected child among their offspring. But, again because human family size is small, there are many pairs of carrier parents who do not have an affected child, so the observed segregation ratio of affected to normal children will be biased upwards from the one to three expected. The segregation ratio calculated in a series of sibships will depend on the method used to collect the data that is to ascertain the families. Methods of ascertainment are listed in Box 5.2. But analysis of segregation ratios only forms part of the investigation once the data have been obtained, and there are other highly informative lines of evidence as well, as shown by the following examples.

Neuromuscular disorders in the Amami islands

The method of study of a rare disorder is well exemplified by the study by Kondo (1992) of Ryukyu progressive spinal muscular atrophy. The Amami islands, at the northern end of the arc that links Kyushu with the island of Formosa (Taiwan), belongs to Kagoshima prefecture. The area of his survey covered all the Amami islands and the northern part of Okinawa island together with the neighbouring islets. The population consisted of all those (287 727) who were alive on the prevalence day (1st October 1965). The target diseases were all those neuromuscular disorders that cause progressive muscular atrophies. Cases were found by the successive information method as follows:

1. Provisional cases were found from the local records for the disabled and for handicapped school children, and from local practitioners.
2. These lists were reviewed with the chief or health officer of each settlement, to whom the target diseases were described, and additional possible cases thus collected.
3. The homes of all suspected cases were visited, the patients were re-examined or, where they were not available, their relatives were asked about their condition.
4. At each home questions were asked about other cases with similar diseases.
5. A final visit ensured that no new provisional cases had been added.

This method was particularly suited for the survey. The settlements are closely knit, endogamous communities, and those individuals who have a disability such as musuclar atrophy are well known to the inhabitants. The data were checked by a subsequent door to door census of all the inhabitants of five settlements chosen at random out of the 56 settlements in Amami-oshima (the largest island of the group), and comparison of the results confirmed the accuracy of the initial collection.

This method brought to light 59 cases of neuromuscular disease, of which 28 were known disorders. But in the remaining 31 cases there was an obscure disease, quite different from any other known neuromuscular disease. Clinical studies of the patients were made and their pedigrees obtained through interviews with the patients and their families. These were checked against the Koseki records, the traditional legal family registers kept in the local municipal offices. Advantage was also taken of the Keizu, a special genealogical record kept in feudal times (1601–1868), intended to maintain the status of a family by clarifying its origin, and many such records are kept by the descendants today.

The case histories, clinical studies, and laboratory investigations (biochemical, histological, electron-microscopic and histochemical) of blood and muscle biopsy specimens conveyed to the base laboratory all showed features unique to the disease. The pedigree study showed that it was a recessively inherited disorder confined to a particular lineage in the pedigrees, and converging on the first Lord of the islands, who was

BOX 5.2.   Methods of ascertainment

Methods of ascertainment were classified by Bailey (1951) and refined by
Morton (1959), and there is a useful illustration of their application in Crow
(1965). The methods may be summarised as follows:

1.   *Complete ascertainment.* Sibships are obtained by random sampling
     of the parents without regard to their children (rarely possible for
     recessive diseases, but possible for occasional recessive serological
     and other normal variants).

2.   *Incomplete ascertainment.* Here families are identified by the
     occurrence of affected children, so that sibships with no children
     affected are omitted. There are three methods:

     a.   *Truncate selection.* Here every affected individual is
          independently ascertained, as would be the case if the
          population were exhaustively examined. A family with a large
          number of affected children is no more likely to be included
          than one with only few. The distribution of affected children
          amongst selected families is approximately binomial except for
          omission of the class with none affected, so the distribution is
          truncated.

     b.   *Single selection.* Here no family is discovered more than once,
          as would be the case when the probability of ascertainment
          approaches zero. This situation might arise for example if
          children of a certain year of age were sampled. The probability
          of a family being included in the study is proportional to the
          number affected, whereas in truncate selection it is independent
          of the number affected.

     c.   *Multiple selection.* In most actual studies, ascertainment lies
          somewhere between truncate and single selection. If the data
          are collected in such a way that the individuals ascertained are
          shown, or if the ascertainment probability is known, there are
          appropriate methods of analysis. Multiple selection includes
          truncate and single selection as special cases.

     Crow (1965) gives an example of a single set of data taken as collected
     by the three methods; the segregation ratios differ, being by multiple
     selection 2.69:1, by single selection 3.29:1, and by truncate selection
     2.24:1. These differing estimates show the dependence of the segre-
     gation ratio on the ascertainment probability. There are several
     methods by which these ratios can be calculated.

References
Bailey, N. T. J. (1951). A classification of methods of ascertainment and analysis in
    estimating the frequencies of recessives in man. *Annals of Eugenics*, **16**, 223–5.

Crow, J. F. (1965). Problems of ascertainment in the analysis of family data. In *Genetics and the Epidemiology of Chronic Diseases*, ed. J. V. Neel, M. W. Shaw & W. J. Schull, pp. 23–44. Washington: United States Department of Health, Education and Welfare.

Morton, N. E. (1959). Genetic tests under incomplete ascertainment. *American Journal of Human Genetics*, 11, 1–16.

appointed when they came under the rule of the Ryukyu principality in the sixteenth century. On account of its unique features and the pedigree evidence, the disease came to be called the Ryukyu type of progressive muscular atrophy.

Multiple sclerosis in the Orkney and Shetland Islands

A similar method, applied in the genetic surveys of multiple sclerosis in Orkney and Shetland, required an amplification, the use of controls. For here the object was different, an attempt to establish the extent of the genetic contribution to the etiology of a complex disorder and this required the possible environmental contribution also to be considered.

Several features of the populations of Orkney and Shetland made them particularly suitable for an attempt at this genetic analysis, for multiple sclerosis there attains its highest reported prevalence. These features include the geographically well-defined area inhabited, the extent and accuracy of the vital records, and the strong sense of identity and the relative constancy of the population. All patients with multiple sclerosis in the Orkney and Shetland Islands alive on 1st December 1974, the prevalence day, were identified from a variety of sources: their case notes in hospitals in Aberdeen, Inverness, Kirkwall and Lerwick; lists drawn up by earlier investigators; and enquiries among the family doctors on the islands. In all cases the diagnosis was reviewed by the investigating team, with further investigations if required. All patients in whom it was confirmed were entered in the survey. For each Orkney or Shetland born patient a contiguous control was selected who was not a member of the same immediate family, who was born in the same parish in the same year, was of the same sex, had lived in the same area for the first 15 years of his life, and was not affected. Also, for the Shetland series, a discontiguous control was selected, matched for age and sex but born and passing his/her childhood in a different parish. Where a control was subsequently found to have a neurological disorder, or to be a first or second degree relation to the patient to whom he/she was matched, a replacement was obtained. Each patient and each control was interviewed to obtain a family history, which usually extended over three of four generations. This was verified from documentary sources and ancestry was then traced further back for each subject to an arbitrary baseline taken as 1775, using the vital records, parish records, electoral returns, land deeds, and the wide variety of documentation that is available for this type of genetic investigation. Blood specimens were obtained from each patient and control and a wide range of gene frequencies compared. The pedigrees of patients and controls were likewise compared and genetic analysis was completed by examination of the inbreeding coefficients of patients and controls, and their kinship coefficients. The inbreeding coefficients appeared to eliminate recessive involvement or rare genes in the etiology, the kinship coefficients eliminated involvement of recently introduced genes dominant or codominant in effect, and the family histories showed that single locus inheritance is unlikely unless penetrance is very low and there is gross environmental interference with gene expression. A multifactorial etiology in which the genetic involvement is polygenic appeared much more likely. Heritability estimates calculated on this hypothesis indicated a genetic contribution to the etiology of the disease in Orkney and Shetland but one that was no more than moderate (Roberts *et al.*, 1979, 1983).

Conclusion

A series of examples illustrates several of the different forms that genetic enquiry in human populations may take. Different populations and different problems require different methods of sampling, data collection, and investigation. There is often information, either inherent in the nature of the population itself, or in documentary form, that will allow the appropriate methods to be chosen and that can be usefully employed in the enquiry. Of essential importance is the initial identification of the problem to be investigated, which guides the design of the enquiry.

Acknowledgements

Acknowledgement is gratefully made to Dr. W. M. G. Tunbridge for permission to use unpublished material from his MD thesis, and to the Multiple Sclerosis Society of Great Britain for financial support.

References

Boorman, K. E. & Dodd, B. E. (1957). *An Introduction to Blood Group Serology*. Churchill: London.

Calderon, R., Wentzel, J. & Roberts, D. F. (1993). HLA frequencies in Basques in Spain and in neighbouring populations. *Annals of Human Biology*, 19, 109–20.

Davies, K. E. (1986). *Human Genetic Diseases: A Practical Approach.* IRL: Oxford.

de Masdevall, M. D., Ercilla, M. C., Arrieta, A. & Vives, J. (1982). Estudio genetico del sistema HLA en la poblacion vasca. *Sangre*, **27**(2), 182–9.

Harris, H. & Hopkinson, D. A. (1976). *Handbook of Enzyme Electrophoresis in Human Genetics.* Amsterdam: North Holland.

Kondo, K. (1992). Inherited neurological diseases in island isolates in southern Japan. In: *Isolation, Migration and Health*, ed. D. F. Roberts, N. Fujiki & K. Torizuka, pp. 130–142. Cambridge: Cambridge University Press.

Mourant, A. E. (1958). Organization for field research. In *The Scope of Physical Anthropology and its Place in Academic Studies*, ed. D. F. Roberts & J. S. Weiner, pp. 25–31. Oxford: Society for the Study of Human Biology.

Mourant, A. E. (1964). Organization for field research. In *Teaching and Research in Human Biology*, ed. G. A. Harrison, pp. 150–60. Oxford: Pergamon.

Mourant, A. E., Kopec, A. C. & Domaniewska-Sobezak, K. (1958). *The ABO Blood Groups.* Oxford: Blackwell.

Roberts, D. F. (1956). A demographic study of a Dinka village. *Human Biology*, **28**, 323–349.

Roberts, D. F. (1986). Who are the Orcadians? *Anth. Anzeiger*, **44**, 93–104.

Roberts, D. F. & Bainbridge, D. (1963) Nilotic physique. *American Journal of Physical Anthropology*, **21**, 341–70.

Roberts, D. F. & Smith, D. A. (1957). A haematological study of some Nilotic peoples of the southern Sudan. *Journal of Tropical Medicine and Hygiene*, 45–52.

Roberts, D. F., Ikin, E. W. & Mourant, A. E. (1955). Blood groups of the northern Nilotes. *Annals of Human Genetics*, **20**, 135–54.

Roberts, D. F., Roberts, M. J. & Poskanzer, D. C. (1979). Genetic analysis of multiple sclerosis in Orkney. *Journal of Epidemiology and Community Health*, **33**, 229–35.

Roberts, D. F., Roberts, M. J. & Poskanzer, D. C. (1983). Genetic analysis of multiple sclerosis in Shetland. *Journal of Epidemiology and Community Health*, **37**, 281–5.

Tunbridge, W. M. G. (1976). The epidemiology of thyroid failure and its relationship to lipid disorders and ischaemic heart disease. MD thesis, University of Cambridge.

Weiner, J. S. & Lourie, J. A. (1969). *Human Biology: A Guide to Field Methods.* Oxford: Blackwell.

6 Nutritional studies in biological anthropology

STANLEY J. ULIJASZEK AND S. S. STRICKLAND

Nutritional studies in biological anthropology are varied in type. They may be directly concerned with nutritional factors and the way that they affect different aspects of human population biology or ecology, or they may be concerned with nutritional adaptation (Haas and Pelletier, 1990). Regardless, some knowledge is needed of nutritional state of either the whole community, or some sector of it. There are various ways in which this can be obtained, and this chapter will be concerned with the description and evaluation of methods that are likely to be of particular interest to anthropologists and human biologists. Details about sampling and analysis of data will not be considered, since these are discussed elsewhere in this book.

A number of volumes have been written about nutritional assessment; most have been concerned with identifying the role of nutritional factors in health problems (Jelliffe, 1966; Mason *et al.*, 1984; Jelliffe and Jelliffe, 1989; Gibson, 1990; Willett, 1990), whilst only two have considered the usefulness of nutritional studies in addressing anthropological problems (Johnston, 1987; Pelto *et al.*, 1989). A good starting point is therefore to consider the types of question that an anthropologist might ask, which would need some understanding of an individual's or population's nutritional state.

Anthropological questions

Nutritional assessment has been defined as the interpretation of information obtained from dietary, biochemical anthropometric and clinical studies, in determining the health status of individuals or populations as influenced by their intake and utilisation of nutrients (Gibson, 1990). For an anthropologist, this is not an end in itself; rather, knowledge of nutritional state based on current beliefs about nutritional assessment can be used to examine aspects of human adaptability and responses to stress under differing physical and cultural circumstances and environments. Further, the existence of nutritionally related health problems in communities and populations can be quantified and statistically analysed, allowing rigorous comparisons to be made between subgroups of a population, or between populations. In this manner, anthropologists have examined nutritional factors as possible explanatory variables in studies of human biology and behaviour.

Amongst the nutrition-related questions that are of interest to biological anthropologists are the following: undernutrition, specific nutrient deficiencies, or obesity in relation to between-population differences in the utilisation of specific nutrients, socio-economic status, modernisation, and subsistence and disease ecology. In addition, there is considerable interest in the relationship between food intake, nutritional status, susceptibility to infectious disease, and the physical growth of children. With such broad areas of interest, the methods of nutritional assessment to be used will depend on the question in hand.

Nutritional assessment

Methods available for nutritional assessment vary in degree of complexity, expense and invasiveness. Some measures are easier or cheaper to make than others; this does not make their use any less sophisticated than more complex or more expensive methods, since the interpretation of nutritional data is always a complex matter.

Nutritional studies carried out by anthropologists can be divided into two main types: those concerned directly with food consumption, and those concerned with nutritional status. There are a number of techniques that can be used:

> Food consumption:
> Twenty-four-hour recall
> Estimated food record
> Weighed food record
> Dietary history
> Food frequency questionnaire

> Nutritional state:
> Anthropometric
> Body composition
> Biochemical
> Immunological
> Clinical

For the first type of study, it may be of interest to know about the types of food eaten by different sectors of a community, whether this differs across seasons, between age-groups, by gender or by reproductive status. Further, if such a study is sufficiently detailed, it may be possible to infer something about nutritional state from weights of foods eaten and the nutrient content of those foods, and by comparing these with recommended daily allowances for nutrients in question. Studies of food and nutrient intakes are usually carried out in association with other types of nutritional assessment, however. For the second type of study, the methods vary in the extent to which they are direct measures of current nutritional state, or indirect measures of past nutritional state. Indirect measures involve the measurement of 'outcome variables', which are the results of past nutrition.

Dietary methods

There is no single, ideal method for gathering dietary intake data. There is no merit in using a more elaborate or expensive method than is needed to obtain the quality of data needed to meet the defined objective of the study. A broad generalisation is that the more detailed the desired data, the more time consuming and expensive is the method needed. No dietary method is free from error, the more detailed methods being subject to types of error different from the less detailed methods. Table 6.1 gives the uses and limitations of the more commonly used methods to assess the food consumption of individuals.

Of these methods, all except the food frequency questionnaire can be used to estimate nutrient intakes. The weighed food method is the most accurate, followed by that of estimated food record. The most appropriate method depends upon the information needed, however (Table 6.2). All of these methods involve errors, and some may affect the food consumption behaviours of the subjects during the period of study. Certainly, the more invasive the method, the more likely is behaviour to be modified.

A number of authors have examined between- and within-subject variation in nutrient intake, in attempts to estimate the number of days necessary to estimate true intakes (Beaton et al., 1979; El Lozy, 1983; Sempos et al., 1985; Willett et al., 1985; Nelson et al., 1989).

Black et al. (1983) have suggested a formula that uses within-subject CV to calculate the number of days of dietary measurement needed to estimate an individual's true intake of a nutrient, with a specified degree of error:

$$D = \frac{r^2}{1 - r^2} \times \frac{Sw^2}{Sb^2} \qquad (6.1)$$

where D = the number of days needed per person, r = the unobservable correlation between the observed and true mean nutrient intakes of individuals over the period of observation, Sw = the observed within-subject coefficient of variation, and Sb = the observed between-subject coefficient of variation.

Equation (6.1) can be used to calculate the minimum number of days of study needed to estimate the intakes of the nutrients of interest, at a chosen level of accuracy. Even if these calculations were carried out *post hoc,* they would be of value for the interpretation of the data collected. For example, if a seven-day weighed dietary intake study had been carried out, and the use of equation (6.1) at $r > 0.9$ showed that the minimum number of days of measured intake needed to estimate energy intake was five, whilst that for vitamin A was 12, then confidence could be expressed for the measurement of intake of the former nutrient but not the latter.

In addition to the problems associated with the use of the methods described in Table 6.1, there are other considerations, such as the sampling method and the need to disaggregate the data by age and sex, which can influence estimates of food and nutrient intake. Further, food preparation techniques can effect the nutrient composition of foods eaten, whilst the non-recording of non-food items of nutritional significance can result in errors of estimate. Table 6.3 gives a summary of the problems associated with these factors.

Anthropologists carry out dietary studies on different units of human organisation, as shown in Figure 6.1. These levels include the individual, the household, band or community subgroup, the community, and the

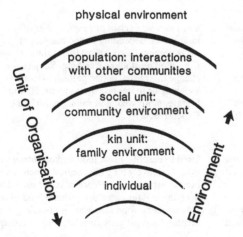

Figure 6.1. Nested character of human organisation and environment (after Foley, 1987).

Table 6.1. *Uses and limitations of commonly used methods to assess the food consumption of individuals*

Method and procedures	Uses and limitations
Twenty-four-hour recall Subject or caretaker recalls food intake of previous twenty-four hours in an interview. Quantities estimated in household measures using food models as memory aids and/or to assist in quantifying portion sizes. Nutrient intakes calculated using food composition data.	Useful for assessing average *usual* intakes of a large population, provided that the sample is truly representative and that the days of the week are adequately represented. Used for international comparisons of relationship of nutrient intakes to health and susceptibility to chronic disease. Inexpensive, easy, quick, with low respondent burden so that compliance is high. Large coverage possible; can be used with illiterate individuals. Element of surprise so less likely to modify eating pattern. Single twenty-four-hour recalls likely to omit foods consumed infrequently. Relies on memory and hence unsatisfactory for the elderly and young children. Multiple replicate twenty-four-hour recalls used to estimate *usual* intakes of individuals.
Estimated food record Record of all food and beverages as eaten (including snacks), over periods from one to seven days. Quantities estimated in household measures. Nutrient intakes calculated using food composition data.	Used to assess *actual* or *usual* intakes of individuals, depending on number of measurement days. Data on *usual* intakes used for diet counselling and statistical analysis involving correlation and regression. Accuracy depends on conscientiousness of subject and ability to estimate quantities. Longer time frames result in a higher respondent burden and a lower co-operation. Subjects must be literate.
Weighed food record All food consumed over defined period is weighed by the subject, caretaker, or assistant. Food samples may be saved individually, or as a composite, for nutrient analysis. Alternatively, nutrient intakes calculated from food composition data.	Used to assess *actual* or *usual* intakes of individuals, depending on the number of measurement days. Accurate but time consuming. Condition must allow weighing. Subjects may change their usual eating pattern to simplify weighing or to impress investigator. Requires literate, motivated, and willing participants. Expensive.
Dietary history Interview method consisting of a twenty-four-hour recall of *actual* intake, plus information on overall *usual* eating pattern, followed by a food frequency questionnaire to verify and clarify	Used to describe *usual* food and/or nutrient intakes over a relatively long time period which can be used to estimate prevalence of inadequate intakes. Such information used for national food policy development, food fortification planning, and to identify food patterns associated with inadequate

Table 6.1 (*cont.*)

Method and procedures	Uses and limitations
initial data. Usual portion sizes recorded in household measures. Nutrient intakes calculated using food composition data.	intakes. Labour intensive, time consuming and results depend on skill of interviewer.
Food frequency questionnaire Uses comprehensive list or list of specific food items to record intakes over a given period (day, week, month, year). Record is obtained by interview, or self-administered questionnaire. Questionnaire can be semiquantitative when subjects asked to quantify usual portion sizes of food items, with or without the use of food models.	Designed to obtain qualitative, descriptive data on *usual* intakes of foods or classes of foods over a long time period. Useful in epidemiological studies for ranking subjects into broad categories of low, medium, and high intakes of specific foods, for components or nutrients, for comparison with the prevalence and/or mortality statistics of a specific disease. Can also identify food patterns associated with inadequate intakes of specific nutrients. Method is rapid with low respondent burden and high response rate but accuracy is lower than other methods.

From Gibson (1990).

Table 6.2. *The choice of method in relation to the information desired from dietary studies*

Desired information	Preferred approach
Actual nutrient intake over finite time period (e.g. in a balance study)	Chemical analysis of duplicate meals *or* calculated intake from weighed records
Estimate of 'usual' nutrient intake in free-living subjects	
Group average	One day intake with large number of subjects and adequate representation of days of week
Proportion of the population 'at risk'	Replicate observations of intake *or* diet history
Individual intake, for correlation or regression analysis	Multiple replicate observations on each individual
Group or individual pattern of food use, proportion of population with particular pattern	Food frequency questionnaire
Average use of a particular food or food group for a group	Food frequency questionnaire *or* a one-day intake with large number of subjects and adequate representation of days of the week

After Beaton (1982).

Table 6.3. *General problems associated with dietary studies*

Sampling	Is the sample of individuals representative of the group or groups under investigation? Can the sample be disaggregated for analysis to allow comparisons between age groups, sexes, households, or any other unit of study? Is the sampling period representative of habitual intakes over longer periods, such as across a year, or season?
Food preparation techniques	Various preparation methods such as peeling, drying, soaking, fermenting, boiling, baking and frying can have significant effects on the nutrient content of foods. The method of preparation should be recorded, to allow reasonable comparison with food composition tables.
Consumption of non-food items	Consumption of some non-food items, such as betel-nut, salt, spices, herbs and water may be of nutritional significance. When carrying out a study in alcohol-consuming groups, care should be taken to ensure that consumption is not under-reported.

Table 6.4. *Food consumption patterns in different populations*

Country	Group	Pattern
Britain, USA, Australia, Canada	Wage-earners and salaried workers	Five-day work-day pattern of food consumption, two-day week-end pattern of consumption. Low seasonal variation in foods eaten; greater seasonal variation in lower socio-economic classes than in higher socio-economic classes.
Papua New Guinea, Senegal, Kenya, Namibia	Groups practising traditional subsistence methods	Seasonal variation in types and amounts of different foods eaten. It is likely that hunter-gatherers and pastoralists experience greater variation than agriculturalists; amongst agriculturalists, it is likely that wealthier households show less seasonal variation than poorer households.
Chad, Cameroun, Gambia	Groups practising traditional subsistence methods	Seasonal post-harvest gorging practised by some, or all, sections of the community.

population. In general, it is better to collect data at lower levels of organisation if at all possible, since such data can be aggregated once collected. Thus, if all members of a sample of households have been included, and collection has been at the individual level, then the data can be aggregated in various ways; the data might be expressed by age group and sex, or it might be aggregated by household, to allow between household comparisons. It is not possible, however, to infer the intakes of individuals from data collected at the household level, because it cannot be assumed that food will be divided either equitably or equally within the household (Wheeler, 1988). The choice of the sampling period is an important consideration in carrying out dietary studies. Human groups exhibit a number of different food consumption patterns; some of these are given in Table 6.4.

Very few groups show no variation in quantity or types of foods eaten either from day to day, or across seasons. Worldwide, patterns of food consumption are almost universally related to the patterns of subsistence. In an industrialised country such as the United Kingdom, there is less variation in types and amounts of food eaten during the usual five-day working period than there is during the two days of the week-end. Indeed, the patterns of consumption may be dramatically different between the two periods, with greater amounts of more expensive foods being eaten on Sunday, and sometimes large amounts of alcohol being consumed on Saturday. In Papua New Guinea, Ningerum horticulturalists show no variation in intake of energy and protein across seasons, but enormous differences in the sources of those nutrients (Ulijaszek, 1985). In Senegal, Serere pastoralists show seasonal variation in energy intake, and in types of foods used (Rosetta, 1988). This is related to the type of agro-pastoralism that these people practise. Mixed horticulturalists in Chad and Cameroun (de Garine and Koppert, 1990), and agriculturalists in the Gambia (Fox, 1953) engage in post-harvest gorging to a greater and lesser extent, respectively. Dugdale and Payne (1986) have shown this practice in the latter group to be a good strategy of food storage.

Seasonal variation in the types of foods eaten does not necessarily mean that energy intake also varies; in some studies it has been shown to be the same (Ulijaszek, 1985). It is more likely that the intake of other nutrients will vary, however, since different foods have different nutrient profiles.

In carrying out a food consumption survey, it is important to obtain some sort of representativeness of the sampling period. For studies in Britain, a sample period of seven days has been suggested, to take account of the difference between work-days and the week-end. Studies using such a period are thus believed to be representative of longer time frames. In developing countries, seasonality may suggest that sampling be carried out at different times of year, representative of the different seasons. Generalising from data that have not taken into account seasonal bias can be dangerous. Lee's (1965) study of the !Kung is instructive in this regard. In carrying out a dietary study at a time of year when the high energy, high protein staple mongongo nut was in season, Lee found that intakes of both these nutrients to be high. Lee's data have been used in support of the 'original affluent society' argument put forward by Sahlins (1972), although it has since been argued that nutritional returns for the !Kung are less good at other times of year, when mongongo nuts are less available (Wilmsen, 1978), and when there is the likelihood of energy nutritional stress (Bentley, 1985).

Without carrying our prior investigations, it is often difficult to determine the most appropriate sampling period. There are various ways of arriving at some approximation. Ethnographic, social and nutritional literature may give some clues as to variation in dietary patterns; alternatively, researchers and government officials who have worked with the population of interest may be able to provide important information. If a group is involved in industrial work, this can give some idea of likely day-to-day variation in food intake, whilst rainfall figures for a district in the tropics can give an indication of the extent of climatic seasonality, which can be translated into seasonality of food intake.

Food preparation techniques can affect the nutrient composition of foods eaten, and these are summarised in Table 6.5. If the intention is to examine nutrient intakes from weighed or recalled dietary data, then care should be taken to note how food is prepared, in order to match the description as closely as possible to those given in the food composition tables. Failure to do so could lead to some gross inaccuracies.

The water content of the food can also affect the calculated nutrient content of the diet, particularly when the consumption of one food supplies the majority of the dietary energy intake. A gross difference in water content between that of the consumed food and that of the published value used to calculate the nutrient intake could lead to enormous inaccuracies in estimated nutrient intakes. For example, 19 samples of sweet potato analysed by Norgan et al. (1979) varied enormously in their energy and protein contents. When corrected for differences in water content, the coefficient of variation between samples was reduced from 39% to 33% for protein content and from 24% to only 2% for energy.

In some populations, the consumption of non-food items could be of

Table 6.5. *The effect of food preparation techniques on the nutrient composition of foods*

Technique	Example	Possible effects
Peeling	Milling of grain; peeling of potatoes and vegetables	Loss of protein and vitamins which are concentrated in the outer layers of the food; greater concentration of carbohydrate per unit of food stuff in the case of cereals
Drying	Drying of fruits, fish, for storage	Loss of ascorbic acid and B vitamins
Pickling or salting	Vegetables, meat	Loss of vitamins; addition of some minerals in large quantities
Soaking	Soaking of cassava to remove cyanide; reconstitution of dried foods	Loss of water-soluble vitamins due to leaching
Sprouting	Sprouting of mung beans, alfalfa	Increase in ascorbic acid content; greater digestibility of protein
Fermenting	Production of tofu, beancurd, shoyu, from soya beans; production of idli from fermented rice and lentil flour	Increased availability of protein
Boiling	Boiling of potatoes and vegetables	Some destruction of vitamins by heat; loss of water-soluble vitamins due to leaching
Baking	Potatoes, tubers in general, meat	Some destruction of vitamins by heat
Frying	Meat, vegetables	Energy content of food greatly increased due to the addition of fat

nutritional significance. In particular, alcoholic beverages contain energy, and in the case of some traditionally prepared drinks, other nutrients too. Subjects consuming even only small amounts of alcoholic drink that are not reported may have their energy intakes significantly under-reported. Another example is the use of betel-nut as a stimulant. Slaked lime (calcium hydroxide) is needed to liberate the narcotic from the nut, and is frequently chewed with the nut. Swallowing only small amounts of this lime in the course of chewing can contribute substantial quantities of calcium to intake; this could easily be missed by the dietary recorder. Further, habituated betel-nut chewers swallow the chewed substance rather than spit it out. Where this happens, betel-nut could also

Table 6.6. *Methods for estimating energy expenditure in largely free-living populations*

Method	Sources of error	Measurement error	Feasible duration	Range of activities	Invasiveness
Intake–balance	Small changes in body composition over intake measurement period	2%	24–360+ h	Total daily EE only	High
Douglas bag or portable respirometer	Leaks and gas diffusion from sampling bags	2–5%	0.1–1.0 h	Moderate to high	Moderate to high
Oxylog	Turbine flow meter	2–5%	0.1–0.5 h	Moderate to high	Moderate to high
Doubly-labelled water ($D_2^{18}O$)	Assumed steady state, known evaporative water losses and isotope fractionation rates, known isotope flux between CO_2, H_2O and other pools	2–6%	72–7200 h	Total daily EE only	Low
Heart rate monitoring	Low or extreme high levels of activity	5–10%	12–72 h	Total daily EE only	Low

EE = energy expenditure.
After Garrow and Blaza (1982).

make an important contribution to protein intake. In studies where trace element intakes are of interest, the consumption of substantial quantities of water, in one form or another, could make important contributions to intakes of calcium or magnesium. In studies of hypertension, the consumption of salt, often disregarded in dietary studies, is of particular interest.

Validation of dietary studies can be carried out by using biochemical markers; these are biochemical indices which give a predictive response to a given dietary component (Bingham, 1984). Examples include twenty-four-hour urinary nitrogen and 3-methyl-histidine excretion as markers of total protein and meat intake, respectively. Although the use of such markers may appear attractive, often single measurements are not adequate to provide accurate estimates of the excretion of the dietary component, however. In addition, standard methods have yet to be widely accepted.

Interpretation of energy intake data will often require estimates of energy expenditure or change in body energy stores. Methods in body composition analysis (see below) enable the latter variable to be studied. In some respects, however, energy studies form a special field in their own right. As with any type of inquiry, the theory of hypothesis to be tested will determine the nature and range of appropriate methods.

Several recent publications review thoroughly the principles underlying techniques of calorimetry (McLean and Tobin, 1987; Blaxter, 1989; Collins and Spurr, 1990) and labelling studies (James, Haggerty and McGraw, 1988; Prentice, 1990). The range of methods likely to be used in human anthropological studies, with an indication of their advantages and limitations, is given in Table 6.6. These methods fall broadly into two groups: those suitable for obtaining estimates of average 24-hour energy expenditure, and those that are appropriate for measuring energy expenditure over periods shorter than one hour.

In general, measurement error falls in the range 2–5%. All methods have been used on free-living populations under non-Western conditions, and have been subject to cross-validation trials by simultaneous measurements on humans, in which the standard of reference has usually been the intake–balance method (Kwalkwarf *et al.*, 1989), the Douglas bag (McNeill *et al.*, 1987; Wenzel *et al.*, 1990), or the whole-body room calorimeter (Seale *et al.*, 1990; Soares *et al.*, 1989).

Equations for predicting basal metabolic rate (BMR) from body weight, with or without other variables, have also been derived. The FAO/WHO/UNU (1985) report, *Energy and Protein Requirements*, advises use of Schofield's (1985) equations. These, however, have been

Table 6.7. *Uses and limitations of commonly used methods to assess the nutritional state of individuals*

Method	Uses and limitations
Anthropometry A combination of measurements including length, height, weight, arm circumference, biceps, triceps, subscapular and suprailiac skinfold thicknesses made on a number of subjects, either cross-sectionally or longitudinally by one or several observers. Comparisons made with standards of variable against age, or accepted cut-offs to judge nutritional state.	Used cross-sectionally in obtaining group estimates of nutritional state, by comparing values for anthropometric measurements with cut-offs obtained from an accepted 'ideal' population. Used longitudinally to assess changes in nutritional state of individuals across time. Limitations include the determination of cut-offs from 'ideal' populations, and defining 'ideal' populations for comparison; some measures are dependent upon knowing the age of children; poor cross-measurement validity of some variables (for example, weight for height and arm circumference). Anthropometry cannot be used to determine the state of a group or individual with respect to specific nutrients.
Body composition Skinfold thicknesses, densitometry, plethysmography, isotope dilution, total body potassium, ultrasound, bioelectrical impedence.	Used to estimate the relative size of different physiological body compartments in order to use them as possible explanatory variables in studies of human functions that are influenced by nutritional state (for example, work output and work capacity, reproductive ecology). Used directly in the study of undernutrition and obesity. All methods apart from skinfold thickness measures and bioelectrical impedence are laboratory-bound.
Biochemistry Measurement of nutrients or their metabolites in biological fluids or tissues, including blood cells and blood cells, hair, fingernails, urine. Measures of changes in blood components or enzyme activities which are dependent on a given nutrient.	Used to obtain direct measures of specific nutrient status. Tests vary in complexity and cost; some can be carried out in the field, whilst others require specialised laboratory procedures. In the latter case, problems of preservation, transport and storage of samples may arise.
Immunology Delayed cutaneous hypersensitivity, secretory IgA, leucocyte antigenic challenge, complement C3.	Used to measure interactions between undernutrition and infection. Tests cannot detect deficits of individual nutrients.

Clinical examination
Physical examination of individual's skin, eyes, hair, mouth, size and shape of parotid and thyroid glands.

Used in identifying extreme pathologies of undernutrition. Some signs are not specific to a deficiency of a single nutrient. Inadequate training in examination techniques or poor standardisation of criteria for judging a sign to be present can lead to inconsistencies in assessments.

From Chandra (1983), Gibson (1990), Jelliffe (1966), Shephard (1991), Weiner and Lourie (1981).

shown to overpredict measured BMR in a wide range of non-Western populations by between 1 and 22% (Henry and Rees, 1991). Thus there is a strong case for recommending direct measurement of BMR in field studies of such groups.

Assessment of nutritional state

A variety of means exists for assessing nutritional state, including anthropometric, body composition, biochemical, immunological and clinical methods. Table 6.7 gives the uses and limitations of methods that are likely to be of use to anthropologists.

Of these, anthropometry, and the measurement of body composition from skinfold thicknesses, are the most widely used methods, because of their cheapness and the portability of the equipment needed. The other methods have been used less often, either because of difficulties of acquiring expertise or laboratory resources, as with clinical and biochemical assessment, respectively, or for logistical reasons, such as the need to transport specimens of body fluids across long distances. Immunological assessment has been little used by either anthropologists or nutritionists, since the methods available are rather new.

Again, the method of choice depends on the question in hand. If a generalised view of the nutritional status of a population, community or group is required, then the measurement of one or more anthropometric variables, such as height, weight or arm circumference, may be adequate. If more subtle effects are to be examined, such as the metabolic changes associated with low protein nutritional status, then one or more biochemical measures might be more appropriate. Immunological tests are likely to be the best measures of functional effects of the interaction between undernutrition and infection. Clinical assessments are only really useful if serious deficiencies of certain nutrients are likely to be found in a substantial proportion of the group being studied.

Anthropometric assessment

Anthropometric assessment is the most widely used because measurement is simple, and, once equipment has been bought, there is no further outlay. Most anthropometric equipment is portable, and measurements can be made in remote geographical regions. Problems associated with anthropometry include measurement error and interpretation of data.

Although nutritional classifications for children exist using anthropometry alone (Waterlow et al., 1977), for adults, a two variable classification has been recommended by James, Ferro-Luzzi and Waterlow (1988) for the assessment of undernutrition. This incorporates body mass

BOX 6.1.   Training in anthropometric measurement

Measurement error exists, even when there is only one trained observer doing all the measuring. Studies in which more than one observer is employed will incorporate both within- and between-observer errors. Since measuring techniques are usually learned by instruction from an expert or supervisor, there is a need to determine when a trainee is ready to perform the measurement accurately. When comparing repeat measurements of a trainee and a trained observer, the differences between the two should be within respectable bounds. Zerfas (1985) has recommended a scheme for evaluating measurement error among trainees. Differences should be consistently below 5 mm for height and arm circumference, 0.1 kg for weight, and 0.9 mm for skinfold thickness before a trainee is deemed as having reached an acceptable level of proficiency. In addition, there should be no systematic bias in measurement between observers. The length of time taken to reach these levels of proficiency varies between trainees; for most, one morning spent in carrying out repeat measurements is enough.

The accuracy of collected anthropometric data can be evaluated by using two error estimates: the technical error of measurement (TEM) and coefficient of reliability (R) (Mueller and Martorell, 1988; Frisancho, 1990). The TEM is obtained by carrying out a number of repeat measurements on the same subject by the same observer, to obtain intra-observer TEM, and (where applicable) the different observers, for inter-observer TEM. Intra-observer TEM, and inter-observer TEM involving only two observers, can be obtained by entering the differences into the equation:

$$\text{TEM} = \surd(\Sigma D^2/2N) \qquad (1)$$

where D is the difference between measurements, and N is the number of individuals measured. If more than two observers are involved, TEM can be calculated using the formula:

$$\text{TEM} = \surd\{\Sigma_1^N[(\Sigma_1^K M(n)^2) - (\Sigma_1^K M(n)^2/K)]/N(K-1)\} \qquad (2)$$

where N is the number of subjects, K is the number of determinations of the variable taken on each subject, and $M(n)$ is the nth replicate of the measurement, where n varies from 1 to K. Acceptable TEMs vary with the measurement taken, and with age (Ulijaszek and Lourie, in press). The table gives maximum acceptable TEMs for height, arm circumference, triceps and subscapular skinfold thicknesses by age group, at two levels of reliability. Ideally, TEM should be lower than the values given for a reliability coefficient of 0.99 (Ulijaszek and Lourie, in press).

(*continued*)

Maximum levels for technical error of measurement at two levels of reliability for either intra- or inter-observer error

Age group (years)	Measurement (males)					Measurement (females)				
	Height (cm)	Sitting height (cm)	Arm circumference (cm)	Triceps skinfold (mm)	Subscapular skinfold (mm)	Height (cm)	Sitting height (cm)	Arm circumference (cm)	Triceps skinfold (mm)	Subscapular skinfold (mm)
Reliability = 0.95										
1–4.9	1.03	0.40(a)	0.31	0.61	0.43	1.04	0.34[a]	0.30	0.65	0.47
5–10.9	1.30	0.35	0.52	0.97	0.87	1.38	0.36	0.54	1.05	1.08
11–17.9	1.69	0.30	0.75	1.45	1.55	1.50	0.29	0.78	1.55	1.74
18–64.9	1.52	0.30	0.73	1.38	1.79	1.39	0.31	0.98	1.94	2.39
65+	1.52	0.30	0.74	1.29	1.74	1.35	0.32	0.98	1.86	2.27
Reliability = 0.99										
1–4.9	0.46	0.18(a)	0.14	0.28	0.19	0.47	0.15[a]	0.13	0.29	0.21
5–10.9	0.58	0.16	0.23	0.43	0.39	0.62	0.16	0.24	0.47	0.48
11–17.9	0.76	0.13	0.33	0.65	0.69	0.67	0.13	0.35	0.69	0.78
18–64.9	0.68	0.13	0.33	0.62	0.80	0.62	0.14	0.44	0.87	1.07
65+	0.68	0.13	0.33	0.58	0.78	0.60	0.14	0.44	0.83	1.02

[a] 2–4.9 years.
From Ulijaszek and Lourie (in press).

References

Frisancho, A. R. (1990). *Anthropometric Standards for the Assessment of Growth and Nutritional Status.* Ann Arbor: University of Michigan Press.

Mueller, W. H. & Martorell, R. (1988). Reliability and accuracy of measurement. In *Anthropometric Standardisation Reference Manual,* ed. T. G. Lohman, A. F. Roche & R. Martorell, pp. 83–6. Champaign, IL: Human Kinetics Books.

Ulijaszek, S. J. & Lourie, J. A. (in press). Intra- and inter-observer error in anthropometric measurement. In *Anthropometry: the Individual and the Population,* ed. S. J. Ulijaszek & C. G. N. Mascie-Taylor. Cambridge: Cambridge University Press.

Zerfas, A. J. (1985). *Checking Continuous Measurements: Manual for Anthropometry.* Division of Epidemiology, School of Public Health. Los Angeles: University of California.

Table 6.8. *Assessment of undernutrition in adults, from body mass index (BMI) and physical activity level (PAL)*

BMI	PAL	Presumptive diagnosis
>18.5	—	Normal
17.0–18.5	>1.4	Normal
	<1.4	CED grade I
16.0–17.0	>1.4	CED grade I
	<1.4	CED grade II
<16.0	—	CED grade III

CED = Chronic energy deficiency, Grad I = mild; II = moderate; III = severe.
From James, Ferro-Luzzi and Waterlow (1988).

index (BMI; weight divided by height squared), and physical activity level (PAL; total daily energy expenditure divided by basal metabolic rate). This is presented in Table 6.8. The authors suggest using energy intake measures to estimate physical activity level, assuming that subjects are in energy balance, and that intake equals expenditure. If this is not available, amendments to this classification may be needed.

Body composition

Body composition measures are useful for estimating body fatness or leanness. The most common method is the use of skinfold measurements to estimate body density from regression equations of skinfolds against body density. Such equations have been derived from studies in which the skinfolds of a large number of subjects have been correlated with their body density as estimated by a 'gold standard' method, such as densitometry or isotope dilution. Although a large number of prediction

---

BOX 6.2.   Calculation of body fatness and fat-free mass from anthropometry

Fat-free mass (FFM) can be calculated in the following manner:

$$FFM = \text{body weight} - [\text{body weight} \times (\%\text{body fat}/100)] \qquad (1)$$

A typical calculation of percentage of body fat and FFM is as follows:

Data: adult New Guinean male, aged 24 years

$$\text{Weight} = 58.4 \text{ kg}$$

Skinfold thickness:

$$\text{Biceps} = 3.4 \text{ mm}$$
$$\text{Triceps} = 5.2 \text{ mm}$$
$$\text{Subscapular} = 8.8 \text{ mm}$$
$$\text{Suprailiac} = 7.3 \text{ mm}$$
$$\text{Total (skinfolds)} = 24.7 \text{ mm}$$

For a 24-year-old male, body density is calculated using the following formula from Durnin and Womersley (1974):

$$\text{Density (g/cm}^2) = 1.1631 - 0.0632 \times \log_{10}[\Sigma(\text{skinfolds})] \qquad (2)$$

$$= 1.1631 - 0.0632 \times \log_{10}(24.7)$$

$$= 1.075 \text{ g/cm}^2$$

$$\text{Fat percentage} = [(4.95/1.075) - 4.5] \times 100$$

$$= 10.47\%$$

$$FFM = \text{body weight} - (\text{body weight} \times (10.47/100))$$

$$= 58.4 - (58.4 \times 0.1047)$$

$$= 52.3 \text{ kg}$$

Such values of fat percentage and FFM may not be typical of Western countries, where individuals other than athletes are generally heavier and fatter than in Papua New Guinea, where body weights are low and fat percentages below 10 are not unusual (Ulijaszek *et al.*, 1989).

References

Durnin, J. V. G. A. & Womersley, J. (1974). Body fat assessed from total density and its estimation from skinfold thickness: measurements on 481 men and women aged from 16 to 72 years. *British Journal of Nutrition*, **32**, 77–97.

Ulijaszek, S. J., Lourie, J. A., Taufa, T. & Pumuye, A. (1989). The Ok Tedi Health and Nutrition Project, Papua New Guinea: adult physique of three populations in the North Fly region. *Annals of Human Biology*, **16**, 61–74.

equations have been developed in this way (Sloan, 1967; Durnin and Womersley, 1974; Jackson and Pollock, 1974; Jackson *et al.*, 1980), only a few have been validated across different populations. The equations of Durnin and Womersley (1974) have been found to be appropriate for Indian men (Jones *et al.*, 1976), Chilean men (Apud *et al.*, 1977), and younger New Guinean adults (Norgan *et al.*, 1982), but not Eskimos (Shephard *et al.*, 1973). Further, the predictive equations derived from measurements of non-pregnant, non-lactating women have been found to be appropriate for lactating women, despite differences in patterns of fat deposition (Butte *et al.*, 1985).

The percentage of body weight as fat can be calculated using one or other of the following equations:

$$\text{Fat percentage} = [(4.950/\text{density}) - 4.5] \times 100 \qquad (6.2)$$

(Siri, 1956)

$$\text{Fat percentage} = [(4.570/\text{density}) - 4.142] \times 100 \qquad (6.3)$$

(Brožek, 1965)

These formulae give good estimates of body fatness, as compared with other methods (Jones and Lourie, 1981), with precision of estimate varying between 3% and 9% when compared with densitometry. The formulae over-estimate the body fatness of women above the age of 60 years by 2–3%, however (Deurenberg *et al.*, 1989). This is because of age-related changes in fat-free body mass, which are greater in females than in males.

Fat-free mass (FFM) consists of skeletal tissue (bone and cartilage), muscle, skin and viscera. Changes in FFM in an individual over time are in large part due to changes in muscle mass, both skeletal and smooth. It can be calculated as shown in Box 6.2.

Another way in which body composition can be estimated in the field is by the use of equations that predict FFM from weight and height measures (Hume and Wyers, 1971), or percentage of fat from BMI (Black *et al.*, 1983). The equations for adults (from Hume and Wyers, 1971) are as follows:

$$\text{Men:} \quad \text{FFM} = 1.39 \times (0.297 \text{ kg} + 0.193 \text{ cm} - 14.01) \qquad (6.4)$$

$$\text{Women: FFM} = 1.39 \times (0.184 \text{ kg} + 0.345 \text{ cm} - 35.27) \qquad (6.5)$$

Here, the calculation in parentheses represents the estimation of total body water; it is assumed that water counts for 0.7194 of FFM, so the reciprocal of this, 1.39, is used to estimate FFM from total body water. The Black *et al.* (1983) equations are:

Men: Fat percentage = $(1.281 \times BMI) - 10.13$ (6.6)

Women: Fat percentage = $(1.481 \times BMI) - 7.0$ (6.7)

A method that has gained recent popularity is the bioelectrical impedence method (BEI), which depends upon the differences in electrical conductivity of fat-free mass and fat. The technique measures the impedence of a weak electrical current ($800 \mu Å$; 50 kHz) passed between the right ankle and right wrist of an individual. The impedence is proportional to the length of the conductor, and indirectly proportional to the cross-sectional area. The length of the conductor is usually a function of the height of the subject. The FFM can be calculated using a formula of the form:

$$FFM = A + (B \times (H^2/R))   (6.8)$$

where A and B are constants, H is height, and R is resistance, or the square of impedence. A number of predictive equations have been developed (Kushner and Schoeller, 1986; Khaled et al., 1988), by relating BEI measures to a 'gold standard' method of estimating body composition, either isotope dilution or densitometry. The applicability of any of these equations to a diversity of populations has yet to be determined, however. Caution is urged in the use of this method, since it has been found that two equations that are used with commercially available instruments under-estimate and over-estimate FFM as determined from an isotope dilution method by an average of 4.7% and 8.1% respectively (Pullicino et al., 1990). In the same study, prediction equations using skinfold thicknesses, weight/height, and body mass index gave underpredictions of only 2.6%, 4.1% and 2.8% respectively. The BEI method may well become an important one for anthropologists in the future, however, when appropriate equations are available.

In addition to these methods, there is a range of laboratory-based techniques, both old and new. These have not been discussed here, since the concern is with field techniques. For further information, the reader is advised to consult the excellent book by Shephard (1991) *Body Composition in Biological Anthropology* and review articles by Lukaski (1987) and Sjostrom (1989).

Biochemical assessment

Although there is a wide range of biochemical tests available for measuring nutritional state (Table 6.9), two are more widely used than any others. These are plasma albumin, which is used to estimate protein nutritional state, and blood haemoglobin, which is used as a measure of iron deficiency, the globally most common mineral deficiency. Before considering the estimation of these two variables, some general aspects of biochemical testing of nutritional state will be considered.

Biochemical tests in the field usually involve either the measurement of a nutrient in biological fluids or tissues, or the measurement of the urinary excretion rate of the nutrient, or a metabolite of it. Tissues and body fluids that have been used include blood, urine, hair, saliva, semen, amniotic fluid, fingernails and skin. The most commonly used of these are urine and blood. It should be noted that the collection of bodily fluids or tissues might seem a bizarre or strange activity in some societies; a person attempting such collections might be suspected of being a sorcerer, and attract hostility as a result. The purpose of the collection should be made very clear to the subjects of the investigation, and no attempt should be made to undertake collection without their consent.

It is clearly impractical to attempt to measure everything, even if it is possible to collect large amounts of sample. In the case of blood, subjects are usually loath to part with more than a few millilitres, if at all. The intended examination of urine allows the collection of large volumes for analysis; it is usually not practical to collect large volumes from each individual because of problems of storage and transport, however. Further, for some analyses, 24 hour urine samples are needed, and there may be poor compliance as a result of the prolonged sampling period. Subjects may be amused or appalled that someone should want them to collect something in a bottle that they might consider unclean.

Hair samples have sometimes been used to determine status of certain trace elements, such as zinc, manganese and iron. These are easy to take, store and transport, but are easily contaminated. The recommendation then is to keep biochemical testing to a minimum; that is, to focus only on the nutrients of specific interest.

There are a number of factors that may confound the interpretation of biochemical tests, so care should be taken to ensure that all samples are collected under standardised conditions. These factors include the following: sample contamination, diurnal variation, physiological state (pregnancy, lactation), infections and inflammations, physical exercise taken prior to measurement, recent dietary intake, recent intake of medicines, present weight gain, present weight loss, and accuracy, precision, sensitivity and specificity of the analytical method. Clearly, the collection, analysis and interpretation of biochemical nutritional information is not a light undertaking. Further information on biochemical measurement can be found in the very useful text on nutritional assessment by Gibson (1990).

Table 6.9. *Biochemical methods for assessing nutritional status*

Nutrient	Method Principal	Supplementary
Protein	Urinary creatinine	Serum insulin-like growth factor 1
	Serum albumin	Serum amino-acid ratio
	Serum transferrin	
	Thyroxine-binding prealbumin	
	Urinary hydroxyproline	
Vitamin A	Serum retinol	
	Serum carotene	
Vitamin B1 (thiamine)	Urinary thiamine	Serum pyruvate
	Erythrocyte transketolase	Serum lactate
Vitamin B2 (riboflavin)	Urinary riboflavin	Erythrocyte riboflavin
	Erythrocyte glutathione reductase	
Niacin	Urinary N′-methylnicotinamide and N′-methyl-2-pyridone-5-carboxylamide	Fasting serum free tryptophan
Vitamin B6	Erythrocyte aminotransferase activities	Tryptophan load test
	Serum pyridoxal-5′-phosphate	Kynurenine load test
	Urinary 4-pyridoxic acid excretion	
Folic acid	Serum folate	Bone marrow morphology
	Erythrocyte folate	
	Haemoglobin	
	Haematocrit	
	Erythrocyte count	
Vitamin B12	Serum B12	Deoxyuridine suppression test
	Erythrocyte B12	Schilling test
		Methylmalonic acid excretion
Vitamin C (ascorbic acid)	Serum ascorbic acid	Urinary ascorbic acid
	Leucocyte ascorbic acid	Salivary ascorbic acid
Vitamin D	Serum 25-hydroxyvitamin D concentrations	Serum calcium and inorganic phosphate
	Serum alkaline phosphatase	
Vitamin E	Serum tocopherol	
	Erythrocyte tocopherol	
	Platelet tocopherol	
	Erythrocyte haemolysis	
	Breath pentane	
Sodium	Urinary sodium	
	Serum sodium	
Iron	Haemoglobin	Bone marrow morphology
	Haematocrit	

Table 6.9 (*cont.*)

| Nutrient | Method | |
	Principal	Supplementary
Iron (*cont.*)	Erythrocyte count	
	Serum iron	
	Total iron-binding capacity	
	Transferrin saturation	
Calcium	Serum calcium	
	Serum ionised calcium	
Iodine	Urinary iodine	
	Serum thyroxine	
	Serum 3,5,3′-triiodothyronine	
Zinc	Serum zinc	Oral zinc tolerance
	Erythrocyte zinc	
	Leucocyte zinc	
	Urinary zine	
	Hair zinc	

From Passmore and Eastwood (1986); Gibson (1990).

Despite reservations, biochemical tests are useful in demonstrating low reserves of nutrients in individuals who do not show clinical signs of specific deficiencies. Samples of urine and blood can easily be collected by field workers if the subjects have no objection, but analyses usually have to be carried out in a distant laboratory. Planning and discussion are therefore needed between field staff and laboratory workers before the study begins, so that transport, storage and quality control are all acceptable, and the study feasible.

For the purposes of nutritional assessment, Gibson (1990) has classified the body's protein stores into two types: somatic, and visceral. Somatic protein is principally composed of skeletal muscle, whilst visceral protein includes the liver, kidneys, pancreas, heart, gastrointestinal tract, serum proteins, erythrocytes, granulocytes and lymphocytes. Somatic protein status has quite a different meaning from visceral protein status, and they are measured by different methods.

Somatic protein status is a measure of skeletal muscle mass, and can be estimated by measuring the excretion of urinary creatinine. Creatinine is derived from the catabolism of creatine phosphate, a high-energy metabolite found mainly in muscle, and its output is significantly related to muscle mass (Cheek, 1968). Somatic protein status of a subject can be estimated using measures of their creatinine excretion in relation to expected levels of excretion. This is the creatinine:height index (CHI), and it is expressed thus:

CHI(%) =
100 × (measured daily excretion/ideal daily excretion for height) (6.9)

Cut-offs for CHI that are suggestive of deficits in body muscle mass are 60–80% of standard (moderate deficit), and less than 60% (severe deficit) (Blackburn et al., 1977).

Visceral protein status can be assessed by measuring total serum protein, serum albumin, transferrin, thyroxine-binding pre-albumin, or retinol-binding protein. Of these, the most commonly used is serum albumin, total serum protein being a rather insensitive measure.

Serum albumin is a reasonable measure of longer-term visceral protein status, since it has a half-life of fourteen to twenty days. Although it has been used as an index of low protein intake, there are various other factors that can affect its level. These include: 1. reduced protein synthesis resulting from inadequate energy intake, or electrolyte, iron, zinc or vitamin A deficiency; 2. altered metabolism due to trauma, stress of infection; and 3. altered distribution of albumin in body fluids in pregnancy (Jeejeebhoy, 1981). Thus, the interpretation of low serum albumin is made difficult; a further difficulty is that serum albumin levels may be elevated in some cases of semi-starvation (James and Hay, 1968), and are elevated in the acute phase of infection. The value of this test is as an index of marginal kwashiorkor, and for the identification of malnourished children susceptible to oedema (Whitehead et al., 1971; Alleyne et al., 1977).

Iron deficiency anaemia is common throughout the world, and there are a number of ways in which this can be estimated in the field. Iron in the body exists as three components: 1. integrally bound to molecules with oxygen-binding or enzymic function; 2. transport iron; 3. storage iron. Usually, levels of the oxygen-carrying molecule haemoglobin are measured as a marker of the exhaustion of iron stores in the liver, and declining levels of circulating iron. There are a number of devices that can be used to determine haemoglobin levels; these vary in cost, accuracy and convenience. A number of such devices are listed in Jelliffe and Jelliffe (1989).

Immunological tests
Tests of immunocompetence cannot be used to detect specific nutrient deficiencies. Rather, they may measure an individual's ability to mount

Table 6.10. *Tests of immunocompetence*

Test	Comments
Lymphocyte count	In malnutrition, the number of lymphocytes is reduced. Levels said to represent undernutrition: 900–1500 cells/mm^3 (moderate); <900 cells/mm^3 (severe).
Thymus-dependent lymphocytes	Reduction in total number of T-cells in under-nutrition. Results expressed as number of T-cells per microlitre of whole blood, and compared with in-house standards for healthy subjects.
Delayed cutaneous-hypersensitivity	Skin test reactivity to specific antigens (purified protein derivative, mumps, trichophyton, *Candida albicans,* dinitrochlorobenzene) is reduced in undernutrition, and when subject is experiencing infectious sepsis.
Secretory IgA	Salivory levels reduced in undernutrition.
Complement C3	Serum C3 is low in undernourished subjects; decreases further with infection.

an immune response to an antigenic challenge; this may be reduced by the combined stresses of nutritional deficiency and infectious disease (Chandra, 1983). Alternatively, they may measure components of the immune system that are impaired by undernutrition. Nearly all aspects of the immune system can be impaired by nutritional deficiency, although only a limited number of tests are currently available. These are summarised in Table 6.10.

Clinical assessment
This method is only useful in detecting the advanced stages of nutritional depletion and it is based on the examination of epithelial tissues, especially the skin, eyes, hair, and mouth, or of organs near the surface of the body, such as the parotid and thyroid glands (Jelliffe, 1966). Although inexpensive, it requires skill. Further, some of the signs used lack specificity. Indeed, most signs of malnutrition are not specific to the lack of one nutrient, and can often be produced by various non-nutritional factors. Thus, care is needed when attempting the clinical assessment of nutritional status.

Details of clinical assessment are given in Jelliffe (1966), Jelliffe and Jellife (1989) and Gibson (1990), and illustrations of various nutritional pathologies are shown in McLaren (1981), Jelliffe (1966) and Jelliffe and Jelliffe (1989).

Standardisation of definition is important to minimise subjectivity in assessment. Researchers intending to carry out such assessments are directed to McLaren's (1981) excellent *Colour Atlas of Nutritional Disorders,* where such definitions are presented in association with colour photographs of the conditions described. If more than one investigator is to carry out the assessment, it is important that the definitions of the signs be available in written form for all investigators. Further, instruction and practical visual training is needed prior to undertaking the work to ensure uniformity of examination technique, and of judgement.

Clinical assessments should not be graded into such categories as 'low', 'medium', 'high', or 0, 1, 2, 3, 4, or −, +, ++, +++, because this increases the level of subjectivity. It is often time-consuming to attempt to differentiate degress of severity of the signs observed. A record of 'positive' or 'negative' is more realistic. An example of the reporting of clinical signs is given in Box 6.3.

BOX 6.3.   Assessment of clinical signs of malnutrition in refugee children from Irian Jaya

In 1985, indigenous people from Irian Jaya, Indonesia, fled across the border into Papua New Guinea. Ulijaszek and Welsby (1985) assessed clinical signs of malnutrition in a number of refugee children at this time.

Age group (years)	Sign	Refugees	Villagers
0–4.9		($N = 155$)	($N = 26$)
	Oedema of wrists and ankles	9	0
	Sparse hair	21	4
	Discoloured hair	15	4
	Moon face	23	0
5–12		($N = 32$)	($N = 13$)
	Oedema of wrists and ankles	0	8
	Sparse hair	9	0
	Discoloured hair	9	0
	Moon face	16	0

The table shows that the extent of clinical nutritional deficiency was greater in the younger age group than in the older one, and that clinical signs were not entirely absent from children in a nearby village, who could be taken as a control group.

Reference
Ulijaszek, S. J. & Welsby, S. M. (1985). A rapid appraisal of the nutritional status of Irian Jaya refugees and Papua New Guineans undergoing severe food shortage in the North Fly region. *Papua New Guinea Medical Journal,* **28**, 42–7.

Conclusion

A number of techniques for the assessment of nutritonal status have been described and in some cases, evaluated. It should be clear to the reader that nutritional assessment is not a simple matter, and that careful thought and preparation should be undertaken prior to undertaking such work. Despite reservations, nutritional assessment techniques form an important part of the battery of methods that human biologists and biological anthropologists use in examining human adaptability and ecology.

References

Alleyne, G. A. O., Hay, R. W., Ricou, D. I., Stanfield, J. P. & Whitehead, R. G. (1977). *Protein–energy Malnutrition.* London: Edward Arnold.

Apud, E., Benavides, R. & Jones, P. R. M. (1977). Application of physiological anthropology to a study of Chilean male forestry workers. *Proceedings of the International Union of Physiological Sciences*, **XIII**, 29.

Beaton, G. H. (1982). What do you think we are measuring? In *Symposium on Dietary Data Collection, Analysis and Significance. Massachusetts Agricultural Experimental Station, College of Food and Natural Resources. Research Bulletin No. 675*, pp. 36–48. Amherst: University of Massachusetts at Amherst.

Beaton, G. H., Milner, J., Corey, V., McGuire, V., Cousins, M., Stewart, E., de Ramos, M., Hewitt, D., Grambsch, P. V., Kassim, N. & Little, J. A. (1979). Sources of variance in 24-hour dietary recall data: Implications for nutrition study design and interpretation. Carbohydrate sources, vitamins and minerals. *American Journal of Clinical Nutrition*, **32**, 2546–9.

Bentley, G. R. (1985). Hunter-gatherer energetics and fertility: a reassessment of the !Kung San. *Human Ecology*, **13**, 79–109.

Black, A. E., Cole, T. J., Wiles, S. J. & White, F. (1983). Daily variation in food intake of infants from 2 to 18 months. *Human Nutrition: Applied Nutrition*, **37A**, 448–58.

Black, A. E., James, W. P. T. & Besser, G. M. (1983). Obesity. A Report of the Royal College of Physicians. *Journal of the Royal College of Physicians of London*, **17**, 5–65.

Blaxter, K. (1989). *Energy Metabolism in Animals and Man.* Cambridge: Cambridge University Press.

Bingham, S. (1984). Biochemical markers of consumption. In *The Dietary Assessment of Populations. Scientific Report No. 4*, pp. 26–30. London: Medical Research Council.

Blackburn, G. L., Bistrian, B. R., Maini, B. S., Schlamm, H. T. & Smith, M. F. (1977). Nutritional and metabolic assessment of the hospitalised patient. *Journal of Parenteral and Enteral Nutrition*, **1**, 11–22.

Brožek, J. (ed.) (1965). *Human Body Composition.* New York: Pergamon Press.

Butte, N. F., Wills, C., Smith, E. O. & Garza, C. (1985). Prediction of body density from skinfold measurements in lactating women. *British Journal of Nutrition*, **53**, 485–9.

Chandra, R. K. (1983). Nutrition, immunity and infection: present knowledge and future directions. *Lancet*, **i**, 688–91.

Cheek, D. B. (1968). *Human Growth: Body Composition, Cell Growth, Energy and Intelligence*. Philadelphia: Lea and Febiger.

Collins, K. J. & Spurr, G. B. (1990). Energy expenditure and habitual activity. In *Handbook of Methods for the Measurement of Work Performance, Physical Fitness and Energy Expenditure in Tropical Populations*, ed. K. J. Collins, pp. 81–90. Paris: International Union of Biological Science.

de Garine, I. & Koppert, S. (1990). Social adaptation to season and uncertainty in food supply. In *Diet and Disease*, ed. G. A. Harrison & J. C. Waterlow, pp. 240–89. Cambridge: Cambridge University Press.

Deurenberg, P., Westrate, J. A. & Kooy, K. (1989). Is an adaptation of Siri's formula for the calculation of body fat percentage from body density in the elderly necessary? *European Journal of Clinical Nutrition*, **43**, 559–68.

Dugdale, A. E. & Payne, P. R. (1986). Modelling seasonal changes in energy balance. In *Proceedings of the XIII International Congress of Nutrition*, ed. T. G. Taylor & N. K. Jenkins, pp. 141–4. London: John Libbey.

Durnin, J. V. G. A. & Wormersley, J. (1974). Body fat assessed from total density and its estimation from skinfold thickness: measurements on 481 men and women aged from 16 to 72 years. *British Journal of Nutrition*, **32**, 77–97.

El Lozy, M. (1983). Dietary variability and its impact on nutritional epidemiology. *Journal of Chronic Diseases*, **36**, 237–49.

FAO/WHO/UNU (1985). *Energy and Protein Requirements. Technical Report Series No. 724*. Geneva: World Health Organisation.

Foley, R. A. (1987). *Another Unique Species. Patterns in Human Evolutionary Ecology*. London: Longman.

Fox, R. H. (1953). A study of the energy expenditure of Africans engaged in various activities, with special reference to some environmental and physiological factors which may influence the efficiency of their work. PhD thesis. London: University of London.

Garrow, J. S. (1981). *Treat Obesity Seriously*. Edinburgh: Churchill Livingstone.

Garrow, J. S. & Blaza, S. (1982). Energy requirements in human beings. In *Human Nutrition*, ed. A. Neuberger & T. H. Jukes, pp. 1–21. Lancaster: MTP Press.

Gibson, R. S. (1990). *Principles of Nutritional Assessment*. Oxford: Oxford University Press.

Haas, J. D. & Pelletier, D. L. (1990). Nutrition and human population biology. In *Human Population Biology*, ed. M. A. Little & J. D. Haas, pp. 152–67. Oxford: Oxford University Press.

Hamill, P. V. V., Drizd, T. A., Johnson, C. L., Reed, R. B. & Roche, A. F. (1977). *NCHS growth curves for children, birth–18 years, United States. DHEW Publication No. (PHS) 78-1650*. Hyattsville, Maryland: National Center for Heath Statistics.

Henry, C. J. K. (1990). Body mass index and the limits of human survival. *European Journal of Clinical Nutrition*, **44**, 329–35.

Henry, C. J. K. & Rees, D. G. (1991). New predictive equations for the estimation of basal metabolic rate in tropical peoples. *European Journal of Clinical Nutrition*, **45**, 177–85.

Hume, R. & Wyers, E. (1971). Relationship between tital body water and surface area in normal and obese subjects. *Journal of Clinical Pathology*, **24**, 235–8.

Jackson, A. S. & Pollock, M. L. (1974). Generalised equations for predicting body density of men. *British Journal of Nutrition*, **40**, 497–504.

Jackson, A.S., Pollock, M. L. & Ward, A. (1980). Generalised equations for predicting body density of women. *Medicine and Science in Sports and Exercise*, **12**, 175–82.

James, W. P. T. & Hay, A. M. (1968). Albumin metabolism: effect of the nutritional state and the dietary protein intake. *Journal of Clinical Investigation*, **47**, 1958–72.

James, W. P. T., Ferro-Luzzi, A. & Waterlow, J. C. (1988). Definition of chronic energy deficiency in adults. *European Journal of Clinical Nutrition*. **42**, 969–81.

James, W. P. T., Haggerty, P. & McGaw, B. A. (1988). Recent progress in studies of energy expenditure: are the new methods providing answers to the old questions? *Proceedings of the Nutrition Society*, **47**, 195–208.

Jeejeebhoy, K. N. (1981). Protein nutrition in clinical practice. *British Medical Bulletin*, **37**, 11–17.

Jelliffe, D. B. (1966). *Assessment of Nutritional Status of the Community. Monograph No. 53*. Geneva: World Health Organisation.

Jelliffe, D. B. & Jelliffe, E. F. P. (1989). *Community Nutritional Assessment. With Special Reference to Less Technically Developed Countries*. Oxford: Oxford University Press.

Johnston, F. E. (ed.) (1987). *Nutritional Anthropology*. New York: Alan R. Liss.

Jones, P. R. M., Bharadwaj, H., Bhatia, M. R. & Malhotra, M. S. (1967). Differences between ethnic groups in the relationships of skinfold thickness to body density. In *Selected Topics in Environmental Biology*, ed. B. Bhatia, G. S. Chhina & B. Singh, pp. 373–6. New Delhi: Interprint Publications.

Jones, P. R. M. & Lourie, J. A. (1981). Fat and lean mass. In *Practical Human Biology*, ed. J. S. Wiener & J. A. Lourie, pp. 87–97. London: Academic Press.

Khaled, M. A., McCutcheon, M. J., Reddy, S., Pearman, P. L., Hunter, G. R. & Weinsier, R. L. (1988). Electrical impedence in assessing human body composition: the BIA method. *American Journal of Clinical Nutrition*, **47**, 789–92.

Kushner, R. & Schoeller, D. A. (1986). Estimation of total body water by bioelectrical impedence. *American Journal of Clinical Nutrition*, **44**, 417–24.

Kwalkwarf, H. J., Haas, J. D., Belko, A. Z., Roach, R. C. & Roe, D. A. (1989). Accuracy of heart-rate monitoring and activity diaries for estimating energy expenditure. *American Journal of Clinical Nutrition*, **49**, 37–43.

Lee, R. B. (1965). Subsistence Ecology of !Kung Bushmen. PhD thesis. Berkeley: University of California.

Lukaski, H. C. (1987). Methods for the assessment of human body composition: traditional and new. *American Journal of Clinical Nutrition*, **46**, 537–56.

Mason, J. B., Habicht, J-P., Tabatabai, H. & Valverde, V. (1984). *Nutritional Surveillance*. Geneva: World Health Organisation.

McLaren, D. S. (1981). *A Colour Atlas of Nutritional Disorders*. London: Wolfe Medical.

McLean, J. A. & Tobin, G. (1987). *Animal and Human Calorimetry.* Cambridge: Cambridge University Press.

McNeill, C., Cox, M. D. & Rivers, J. P. W. (1987). The Oxylog oxygen consumption meter: a portable device for measurement of energy expenditure. *American Journal of Clinical Nutrition*, 45, 1416–19.

Nelson, M., Black, A. E., Morris, J. A. & Cole, T. J. (1989). Between- and within-subject variation in nutrient intake from infancy to old age: estimating the number of days required to rank dietary intakes with desired precision. *American Journal of Clinical Nutrition*, 50, 155–67.

Norgan, N. G., Durnin, J. V. G. A. & Ferro-Luzzi, A. (1979). The composition of some New Guinea Foods. *Papua New Guinea Agricultural Journal*, 30, 25–39.

Norgan, N. G., Ferro-Luzzi, A. & Durnin, J. V. G. A. (1982). The body composition of New Guinean adults in contrasting environments. *Annals of Human Biology*, 9, 343–53.

Passmore, R. & Eastwood, M. A. (1986). *Human Nutrition and Dietetics.* London: Churchill-Livingstone.

Pelto, G. H., Pelto, P. J. & Messer, E. (1989). *Research Methods in Nutritional Anthropology.* Tokyo: United Nations University.

Pilch, S. SM. & Senti, F. R. (eds.) (1984). *Assessment of the Iron Nutritional Status of the US Population Based on Data Collected in the Second National Health and Nutrition Examination Survey, 1976–1980.* Bethesda, Maryland: Federation of the American Societies for Experimental Biology.

Prentice, A. M. (1988). Applications of the doubly-labelled water ($^2H_2{}^{18}O$) method in free-living adults. *Proceedings of the Nutrition Society*, 47, 259–68.

Prentice, A. M. (1990). Long-term energy expenditure measurement: stable isotope method. In *Handbook of Methods for the Measurement of Work Performance, Physical Fitness and Energy Expenditure in Tropical Populations*, ed. K. J. Collins, pp. 91–4. Paris: International Union of Biological Sciences.

Pullicino, E., Coward, W. A., Stubbs, R. J. & Elia, M. (1990). Bedside and field methods for assessing body composition: comparison with the deuterium dilution technique. *European Journal of Clinical Nutrition*, 44, 753–62.

Rosetta, L. (1988). Seasonal variations in food consumption by Serere families in Senegal. *Ecology of Food and Nutrition*, 20, 275–86.

Sahlins, M. (1972). *Stone Age Economics.* London: Routledge.

Sauberlich, H. E., Dowdy R. P. & Skala, J. H. (1974). *Laboratory Tests for the Assessment of Nutritional Status.* Cleveland, Ohio: CRC Press Inc.

Schofield, W. N. (1985). Predicting basal metabolic rate, new standards and review of previous work. *Human Nutrition: Clinical Nutrition*, 39C, Supplement 1, 5–41.

Seale, J. L., Rumpler, W. V., Conway, J. M. & Miles, C. W. (1990). Comparison of doubly-labelled water, intake-balance and direct- and indirect-calorimetry methods for measuring energy expenditure in adult men. *American Journal of Clinical Nutrition*, 52, 66–71.

Sempos, C. T., Johnson, N. E., Smith, E. L. and Gilligan, C. (1985). Effects of intraindividual and interindividual variation in repeated dietary records. *American Journal of Epidemiology*, 121, 120–30.

Shephard, R. J. (1991). *Body Composition in Biological Anthropology.* Cambridge: Cambridge University Press.

Shephard, R. J., Hatcher, J. & Rode, A. (1973). On the body composition of the Eskimo. *European Journal of Applied Physiology*, **32**, 3–15.

Siri, W. E. (1956). Gross composition of the body. *Advances in Biological and Medical Physics*, **4**, 239–80.

Sjostrom, L. (1989). Recent methods in the study of body composition. In *Auxology 88. Perspectives in the Science of Growth and Development*, ed. J. M. Tanner, pp. 353–66. London: John Libbey.

Sloan, A. W. (1967). Estimation of body fat in young men. *Journal of Applied Physiology*, **23**, 311–15.

Soares, M. J., Sheela, M. L., Kurpad, A. V., Kulkarni, R. N. & Shetty, P. S. (1989). The influence of different methods in basal metabolic rate measurements in human subjects. *American Journal of Clinical Nutrition*, **50**, 731–6.

Ulijaszek, S. J. (1985). Seasonal variation of food intake in Ningerum villagers of Papua New Guinea. Poster presented at the XIII International Congress of Nutrition, Brighton, UK, 18–23 August.

Ulijaszek, S. J. (1990). Commentary on 'Towards an Integrated Medical Anthropology'. *Medical Anthropology Quarterly*, **4**, 374–9.

Waterlow, J. C., Buzina, R., Keller, W., Lane, J. M., Nichaman, M. Z. & Tanner, J. M. (1977). The presentation and use of height and weight data for comparing the nutritional status of groups of children under the age of 10 years. *Bulletin of the World Health Organization*, **55**, 489–98.

Weiner, J. A. & Lourie, J. A. (1981). *Practical Human Biology.* London: Academic Press.

Wenzel, C., Wenzel, H. G., Golka, K., Rutenfranz, M. & Rutenfranz, J. (1990). A comparitive study on different methods for the determination of energy expenditure. *International Archive on Occupational and Environmental Health*, **62**, 101–3.

Wheeler, E. (1988). *Within-Household Food Allocation. Department of Human Nutrition Occasional Paper No. 8.* London: London School of Hygiene and Tropical Medicine.

Whitehead, R. G., Frood, J. D. L. & Poskitt, E. M. E. (1971). Value of serum-albumin measurements in nutritional surveys. A reappraisal. *Lancet*, **ii**, 287–9.

Willett, W. C. (1990). *Nutritional Epidemiology.* Oxford University Press.

Willett, W. C., Sampson, L., Stampfer, M. J., Rosner, B., Bain, C., Witschi, J., Hennekens, C. H. & Speizer, F. E. (1985). Reproducibility and validity of a semiquantitative food frequency questionnaire. *American Journal of Epidemiology*, **122**, 51–65.

Wilmsen, E. (1978). Seasonal effects of dietary intake on Kalahari San. *Federation Proceedings*, **37**, 65–72.

WHO (1972). *Nutritional Anaemia. Technical Report Series No. 3.* Geneva: World Health Organisation.

7 Historical demography and population structure

JAMES H. MIELKE AND ALAN C. SWEDLUND

The inclusion of demographic data in anthropological studies has a long, but spotty past. Early works were primarily descriptive and non-theoretical in nature. However, some of these early studies were insightful, and the issues explored are still researched and debated (e.g., see Carr-Saunders, 1922; Firther, 1936). Demographic anthropology did not really develop until the 1970s, and according to Baker and Sanders (1972) the impetus for this interest and growth was threefold: 1. studies of nonhuman primates in their natural environments during the 1960s spawned a body of descriptive data that lent itself to demographic analyses; 2. there was a growing realization among anthropologists that demographic variables were integral to understanding human population genetics (promoted earlier by Lasker, 1954, and Spuhler, 1959, for instance); and 3. there was increasing interest in human ecology and population differences.

The incorporation of demographic concepts and approaches in anthropological studies has served many purposes. Hammel and Howell (1987) list four observations that influence their advocacy for the inclusion of population theories in anthropology. Firstly, all human societies assign moral and emotional significance to basic demographic events such as birth, migration, marriage, and death. Secondly, the quality of research and innovations that have characterized recent work in anthropological demography is encouraging. Thirdly, demography crosscuts and unifies anthropology unlike any other subject because it can be the basis for observation, discussion, integration, hypothesis testing, and collaboration among the subdisciplines of anthropology. Weiss (1976, p. 351) shares this same view: 'various aspects of demography have been shown to be immensely useful in developing this [anthropological] theoretical foundation, and it has become clear that demographic variables can form a unifying metric for anthropology'. Fourthly, a demographic perspective is important in anthropological theory building and in the integration of cultural and biological evolution.

Reviews of the state of demography within anthropology have often

140

examined the changing attitudes and approaches of anthropologists to the incorporation of demographic data or theories into their research (e.g. Howell, 1973, 1986; Petersen, 1975; Weiss, 1975, 1976, 1989; Swedlund, 1980; Caldwell *et al.*, 1987; Leslie and Gage, 1989). Even with all of the research and interest in demography, Hammel and Howell (1987) and Weiss (1989) state that we still do not have a theory of anthropological demography (or a theory of the relationship between sociocultural evolution and population) that can be built upon in an additive manner. Hammel and Howell advocate that we place anthropological demography within the framework of modern evolutionary theory.

While these studies point to the fact that a consensus regarding the role of demography in anthropology has not yet clearly emerged, they also point to the acceptance and increasing implementation of demographic approaches in anthropological inquiry. In biological anthropology the significance of demographic theories and data has never been seriously questioned, but it was the interest in population structure following the pioneering work of Sewell Wright that stimulated a flurry of work in genetics and demography during the 1950s and 1960s.

When attempts were undertaken to test empirically the elegant theories being developed in population genetics, it was quickly recognized that the human populations typically studied by anthropologists often lacked an historical record with which generational changes could be measured. This, in turn, stimulated research into historical population structure. The purpose of this chapter is to introduce this area of inquiry, to review concepts and basic methods, and to provide examples of the outcomes from some selected studies. Finally, it will also briefly outline those new directions in research that appear to be focused on a broader range of anthropological questions than those initially envisioned.

Historical population structure

Demographers utilize the concept of population structure to mean the sex and age composition of a particular population. This perspective also takes into consideration births, deaths, and migration that occur over time. Population structure from a genetic standpoint, and as employed in this paper, has been defined as all the consequences of the mechanisms other than differential selection and mutation affecting change in the gene or genotype frequencies in a population (Cavalli-Sforza, 1959). Yasuda and Morton (1967) define population structure as the totality of deviations from panmixia or random mating. Thus, population structure includes the effects of inbreeding (consanguinity), assortative mating, random genetic drift, and migration, and takes into account such factors as the size and distribution of subdivided populations (Lasker, 1954; Morton, 1969; Yasuda and Morton, 1967; Harrison and Boyce, 1972; Cannings and Cavalli-Sforza, 1973; Carmelli and Cavalli-Sforza, 1976). Even though natural selection is not technically considered part of population structure, this chapter will also explore the uses of historical data in examining selection in human populations.

If one is interested in examining changes in population structure over time, it is only logical to utilize historical data sources. Historical populations provide the researcher with the opportunity to investigate changes in the rates of inbreeding, migration, and genetic drift; to estimate their effects; and to test the robustness of models employing assumptions of constant conditions. Although there are a few early studies of population structure that employed historical data (e.g., Sutter and Tabah, 1955), major contributions started to appear in the mid-1960s (Cavalli-Sforza, 1962; Alström and Lindelius, 1966; Küchemann et al., 1967; Cavalli-Sforza and Zei, 1967; Roberts, 1968). Since that time numerous studies have been done in Western Europe (e.g., Cavalli-Sforza, 1969; Hiorns et al., 1969; Hussells, 1969; Morton and Hussells, 1970; Dobson and Roberts, 1971; Roberts and Sunderland, 1973; Roberts and Rawlings, 1974; Ellis and Starmer, 1978; Colman, 1981; Lees and Relethford, 1982), Fenno-scandia (e.g., Beckman and Ceder-gren, 1971; Mielke et al., 1976, 1982; Workman et al., 1976; Kramar, 1979; Jorde et al., 1982), Japan (e.g., Yasuda and Kimura, 1973; Yasuda, 1975), the Americas (e.g., Steinberg et al., 1967; Halberstein and Crawford, 1972; Swedlund, 1972; Morgan, 1973; Crawford, 1976), and elsewhere. As will be seen in the following pages, research has continued in these and many other areas with increasing resolution of many of the questions initially asked and new insights into the ways in which culture and human behavior affect panmixia and microevolutionary processes. We have also gained an understanding of the ways in which humans reflect patterns observed in mammalian population ecology and the ways in which they do not.

Combining the basic data of births, marriages and deaths with other information enriches the picture. For example, information on health or cause of death, socioeconomic status, residential location, ethnicity and many other variables are obtainable in a variety of historical settings; but, before describing the methods and results of this research, an under-standing of the fundamentals is essential: What are the sources and types of historical data, and what processes do they permit one to estimate? A conservative approach is taken in this discussion, and the reader is cautioned on the importance of selecting data sources carefully and of pursuing data collection in a very systematic way. The data collection phase can be a tedious process, but the results will only be as good as the information on which they are based.

Historical demographic data sources

Historical demographic data can be obtained from a myriad of sources. Hollingsworth (1969, pp. 43–44) lists, 'in relative order of usefulness', 19 sources:

1.   censuses, especially if given by name and age
2.   vital registration data
3.   Bills of Mortality
4.   ecclesiastical records, such as parish registers and communicants' lists
5.   fiscal documents
6.   military records
7.   inventories of property
8.   genealogies
9.   wills
10. marriage settlements
11. eye-witness estimates
12. prices, over the long term
13. number and extent of towns
14. archaeological remains
15. methods of agricultural economy
16. ecclesiastical and administrative geography
17. new buildings
18. colonization of new land
19. cemetery data, both from skeletons and tombstone inscriptions

Probably the best discussion of most of these types of data sources and their reliability can be found in Willigan and Lynch (1982). Their data classification scheme will be the basic form used in this discussion.

Parish registers

The most widely used sources in historical demography consist of parish registers and civil registration records. These sources have become the mainstay for European demography since the early work of Fleury and Henry (1958) for several reasons: they provide continuous demographic information on individuals; the data can often be linked using family

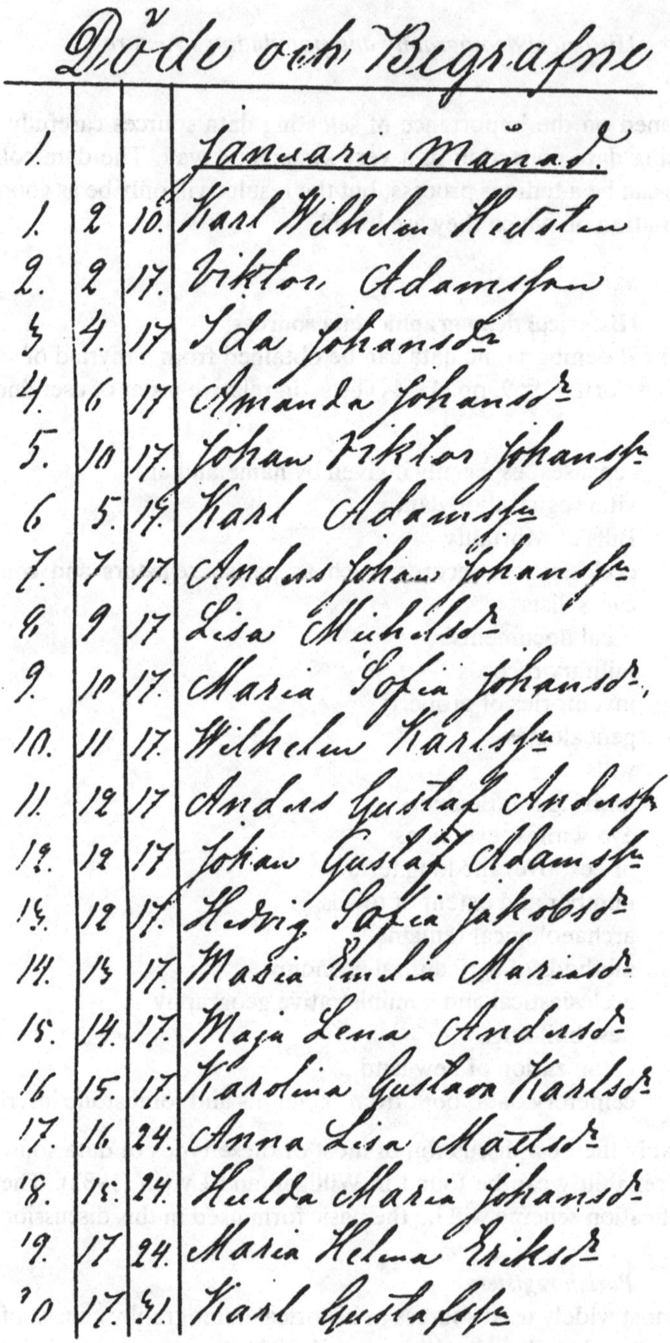

1. Januari Månad.

1.	2	10.	Karl Wilhelm Hedström
2.	2	17.	Viktor Adamsson
3.	4	17.	Ida Johansdr
4.	6	17.	Amanda Johansdr
5.	10	17.	Johan Viktor Johansson
6.	5	17.	Karl Adamsson
7.	7	17.	Anders Johan Johansson
8.	9	17.	Lisa Michelsdr
9.	10	17.	Maria Sofia Johansdr
10.	11	17.	Wilhelm Karlsson
11.	12	17.	Anders Gustaf Andersson
12.	12	17.	Johan Gustaf Adamsson
13.	12	17.	Hedvig Sofia Jakobsdr
14.	13	17.	Maria Emilia Mariasdr
15.	14	17.	Maja Lena Andersdr
16.	15	17.	Karolina Gustava Karlsdr
17.	16	24.	Anna Lisa Mattsdr
18.	15	24.	Hulda Maria Johansdr
19.	17	24.	Maria Helena Eriksdr
20.	17	31.	Karl Gustafsson

Figure 7.1. Example of a death record (*Deaths and Burials – year – 1869*) from Finland. Column 1: numerical order of deaths; Column 2: date of death; Column 3: burial date; Column 4: month is indicated (*January*) followed by

(*Continued opposite.*)

år — 1869.

Trp. barn	4	vangiven	Tyhäniemi
d:o	5	Kikhosta	Okerois Kauppila
Eb. barn	11	d:o	Kahmajärvi Tapu..
d:o	4	d:o	d:o
d:o	6	d:o	d:o
gift Tjust	56	Horssbra	Stein..da Appola
Tjust gift	40	d:o	Okerois Tankkala
Bhu	31	Lungsot	Lahti's Mäkelä
Tjust ogift	20	Typhus	Okerois Tankkala
Trp. barn	2	vattsot	Prästgårds Tellot
Spl. ogift	35	Lungsot	Lahti's Luikkola
d:o gift	28	Typhus	Okerois Tankkala
Bf. gift	80	ålderdom	Kastari Kerola
oä.bl. barn	4	Kikhosta	Sairakkala Sikosta
Trp. d:o	51	Typhus	Okerois Tankkala
Spl. barn	8	Kikhosta	Mukkula gård
Bf. Eko	70	ålderdom	Näsäkkä Kandola
Trp. barn	2	Sm: koppor	Tyhäniemi Mukku..
Inh: h:o	44	Typhus	d:o Ylöstalo
Bt gift	72	ålderdom	Okerois Ayala..

the names of the deceased; Column 5: occupation/social class and/or civil
status (e.g., married, widow, wife, child, unmarried) of deceased; Column 6:
age at death; Column 7: cause of death; Column 8: place of residence.

reconstitution techniques (Fuster, 1986; Hinde, 1987; Bocquet-Appel and Jakobi, 1990); life history analyses may be conducted; and aggregate level statistics can be compiled, providing information on such things as the incidence and intensity of epidemics, famines, and war (Mielke *et al.*, 1984; Bittles *et al.*, 1986; Mielke and Pitkänen, 1989; Smith *et al.*, 1990). Parish records (and later civil records or registrations) are usually similar in structure and form, providing information on named individuals. The registers were most often started by local clergy in order to comply with directives (e.g., eccliastical law) issued by the church (state) to keep track of the inhabitants within a region. The contents of parish registers vary; however, they basically provide information on baptisms (births), marriages, and burials (deaths). Space does not permit a detailed examination of each of these three records. However, an example of a relatively 'ideal' death record would contain the following types of information: date of death, date of burial, name of deceased, sex, occupation or social class, residence at time of death, age at death (or date of birth), marital status, and cause of death (see Figure 7.1).

Obviously, the entries in the records need to be assessed for quality and accuracy. Ideally, there should be few or no temporal gaps in the data set. Quality assessment can take numerous forms depending upon the type of record (see e.g., Gutman, 1956; Henry, 1968; Drake, 1974; Swedlund *et al.*, 1976, 1980; Lee, 1977; Pitkänen, 1977; Wrigley, 1977; Wrigley and Schofield, 1981; Mielke and Pitkänen, 1989; Smith and Pain, 1989).

Censuses and enumerations

There are minor technical differences between censuses and enumerations (Willigan and Lynch, 1982, p. 79), with the term census usually being reserved for a written compilation of all inhabitants in a specific, well-defined region. Enumerations are often limited to a specific group or category of people (when not everyone was of interest to the state or government) and are usually pre-nineteenth century. A basic unit of

---

Figure 7.2. Example of a census/enumeration (*Form for Reporting the Population in the Countryside*) from Finland. The heading reads: *Year 1810 Population in all the Parishes in Tavastehus deanery Åbo Dioceses and Tavastehus Province*. The sections are: I.: *Part, According to the Age* then below that is the *Age* and a column for *Males* and one for *Females*. II.: *Part, According to Marital Status* (e.g., married, widowed). III.: *Part, According to Comparison* (this section gives the population count from five years ago and then the intervening number of births and deaths – the difference between the figures, 21 217 and 21 101 is mainly due to migration). IV.: *Part, According to Status and Circumstances* ('*Occupation*') (this section continues for three pages).

FORMULÄR
förFolkmängdens antecknande på Landsbyggden.

År *1810* Folkmångd uti *alla* Församlingar
Tavastehus Prosteri *660* Stift och *Tavastehus* Län.

I. Art. Efter Åldren.			IV. Art. Efter Stånd och Vilkor.		
Ålder.	Mank.	Qvink.	**A. Mankönet.**		
Under 1 År	335	373	**§. a.**		**§. f.**
Mellan 1 och 3 . . .	388	381	Präster	47	Bönder på egna Hemman
- 3 — 5 . . .	531	521	Lärare vid Skolar . .		Bönder på andras Hemman
- 5 — 10 . .	1332	1364	Öfnings-Mästare . . .		Torpare . . . . .
- 10 — 15 . . .	1179	1165	Studerande	8	Nybyggare sedan sista Qvin-
- 15 — 20 . . .	991	1051	Kyrko-Betjänter . . .	21	qvennium . . . .
- 20 — 25 . . .	936	826			Arbetsföre Backstugu Boer
- 25 — 30 . . .	793	773	Summa	49	Arbetsföre Inhyses Män .
- 30 — 35 . . .	746	832			Skär-Bönder och Fiskare
- 35 — 40 . . .	635	628	**§. b.**		Åldrige och bräcklige Bönder
- 40 — 45 . . .	678	633	Civile Ämbets- och Tjänstmän		och Torpare, som upphört
- 45 — 50 . . .	481	583	högre och lägre . .	9	med Landtbruket . . 304
- 50 — 55 . . .	418	431	Medici		Bondedrängar 2326
- 55 — 60 . . .	350	433	Fältskärer		— Gossar 12
- 60 — 65 . . .	262	297	Kron-Betjäning . . .	6	
- 65 — 70 . . .	162	179	Tull-Betjäning . . .		Summa 5394
- 70 — 75 . . .	82	127	Bergs-Betjäning . . .		
- 75 — 80 . . .	12	63	Jägeri- och Skogs-Betjäning	2	**§. g.** Mästare. Lärling.
- 80 — 85 . . .	14	23	Sluss- Bro- och Färje-Betjäning	2	Byggmästare . . . 1
- 85 — 90 . . .	1		Dykeri-Betjäning . . .		Murare 3
- 90 — 95 . . .			Summa	19	Målare . . . .
- 95 — 100 . . .	-	1			Orgel-Byggare . . .
101, 102, 103 &c. .	—	—	**§. c.**		Sadelmakare . . .
	—	—	Officerare		Skomakare . . . 36 31
	—	—	Under-Officerare . .		Skräddare . . . 48 30
	—	—	Soldater och Båtsmän . .		Smeder och Hofslagare 42 10
	—	—	Skepps- och Fält-Timmermän		Snickare . . . 2 2
Summa	10374	10895	Musikanter och Trumslagare .		Stenhuggare . . .
A. Hela Summan	21217		Träss-Kuskar och Drängar .		Stolmakare . . .
					Svarfvare . . . 1
II. Art. Efter Gift och			Summa		Tunnbindare . . .
Ogift Stånd.			**§. d.**		Urmakare . . .
Gifte	3639	3659	Skeppare		Vagnmakare . . . 1
Enklingar och Enkor . .	743	859	Sjömän		
Ogifte öfver 15 år . .	8419	2939	Lotsar		
Ungdom under 15 år . .	3967	3952	Fyrbåks-Vaktare . .		
Summa	10374	10895	Summa		
B. Hela Summan	21217		**§. e.**		Summa 139 51
			Utur tjänst gångne af §. a.	1	
III. Art. Efter jämförelse.			— — b. .		**§. h.**
			— — c.	12	Privates Fogdar . . 12
Summan af Folkhopen i före-			— — d. .		Kammartjänare . . 2
gående Quinquennii Tabell .	10071 10623		— — g. .	4	Trägårdsmästare . . 2
Flere (Flere) Födde än Döde			— — h. .		Trägårdsdrängar . . 1
alla 5 åren sedermera . .	14 193		— — h. .		Lakäjer och Uppassare 1
			Possessionater utan Titel och		Jägare . . . .
Summa eller Skillnad	10085 11016		Tjänst	1	Skogvaktare . . . 3
C. Hela Summan	21101		Utflyttade Borgare ifrån Städerna		Kuskar . . . .
			Borgare uti Köpingar och Man-		Fiskare . . . .
			talsskrifne på Landet		Gårdsdrängar . . 66
			Landtmän, som icke kunna hän-		Tjänstgåssar . . 15
			föras till annan Titel . .	1	
			Summa	91	Summa 104

enumeration was frequently the household or family (see Cook and Borah, 1971). At other times the resultant record was a simple population total, sometimes broken down by age, sex, and/or other categories of interest (see Figure 7.2). Regular censuses evolved in Europe and the Americas during the nineteenth century and take a number of forms, often listing family and household composition, ages, occupations, and other sociocultural measures of interest to the demographer and anthropologist.

Again, one should assess the quality and accuracy of enumerations and censuses (see e.g., Wrigley, 1972; Weitman et al., 1976; Lawton, 1978; Wrightson and Levine, 1979).

Genealogies

Genealogies are valuable sources for studying individuals, families (e.g., nobility), and sometimes whole populations from a longitudinal perspective. Genealogies can also be considered as reconstituted families and lend themselves nicely to cohort analysis. When using genealogies one must be aware of potential biases such as non-biological links and kin, falsification, removal of select individuals (e.g., illegitimate children, unsavory characters such as thieves and scoundrels, and out-migrants who did not write 'home'), and fictitious links.

Anthropological studies employing genealogies are numerous, and only a few studies that illustrate different approaches or points will be mentioned (see e.g., Dyke and Morrill, 1980). A study by Adams and Kasakoff (1980), which compared marriage migrations rates in Colonial New England computed from genealogies with rates obtained from vital records, clearly demonstrates the problems one may encounter using genealogies alone. These problems should be considered if one is going to use genealogies. Genealogies can also be used to examine inbreeding, isonymy, and consanguinity (e.g., Roberts, 1971; L. A. Rogers, 1987; O'Brien et al., 1988a, 1988b; Jorde, 1989; Bocquet-Appel and Jackobi, 1990). See Darlu and Cazes (1988) for a cautionary note on possible bias.

Population registers

Population registers combine some characteristics of both genealogies and censuses (enumerations), that is, the data cover a specified population, as does a census, but also involve named individuals and families (households), similar to genealogies. According to Willigan and Lynch (1982) there are two major types of population registers: 1. Religious registers such as the *shūmon aratamē chō* of Japan, and 2. European registers, the best known being the Swedish *husförhörslängd* (General

Parish Register or Parish Main Books). These records can be used for a variety of studies; and when coupled or linked with other data sets (such as birth, marriage, death, and vaccination records) can provide a vast amount of interesting data for the anthropological demographer. These types of records have been used to study such things as twinning, fertility, household and family composition, and the opportunity for natural selection (Eriksson, 1973; Åckerman *et al.*, 1978; Devor, 1979; Hed, 1981, 1984, 1986b; Rogers and Norman, 1985; Trapp, 1987).

Organizational and institutional records
Many written records kept by organizations or institutions on groups of people or individuals can be considered sources of interest to the demographic anthropologist. Information recorded by these groups may be directly useful or provide supplementary data. Such groups usually share characteristics and are thus considered as selective samples (or subpopulations) of the population at large. Examples of organizational records include those kept by courts, the police, and poor-relief agencies. Records of institutions, usually defined as formal organizations that contain a resident population, such as prisons, asylums, monasteries, and the military, may contain supplemental information not available in other historical sources. For further discussion of these types of records, see Willigan and Lynch (1982).

Limitations of the data
The use of historical demographic data in examining population (genetic) structure has limitations, some inherent in the data sets themselves. Firstly, the most evident limitation for studying the genetic structure is the lack of information on the genes. Little, if anything, can be done to correct this basic problem. It may, however, be possible to bioassay the contemporary population and then use the population history to account for the observed genetic variation. This approach has been used success-fully in many studies and has been one of the more productive appli-cations of historical data. Obviously, this approach is limited in scope, and the findings are specific to the population under consideration. In fact, Harpending (1974, p. 229) concludes that 'studies of the genetic structure of small populations have made particular and incidental contributions to formal genetics, to regional history and prehistory, to epidemiology, and to several other fields to which they are peripheral, but that they have not advanced our understanding of human evolution in a global sense'. This approach is also restrictive and may eliminate many potential data bases from consideration. Therefore, it is often desirable

to consider historical data as a reflection of the evolutionary changes and processes that have occurred in the shaping of the genetic structure of a population.

Secondly, the demographic records were not collected with the historical demographer or anthropologist in mind. The purposes were usually political, economic, and/or ecclesiastical. In some cases there may have been medical (biological) reasons for record keeping, but this is a rarity. In many instances this limitation may be reflected in the types of research questions that can be examined given the particular historical data set. While some researchers may consider this limitation a major factor, other historical demographers (e.g., van der Walle, 1976) argue that using data for purposes other than those for which they were collected is appropriate since the original respondents and record keepers would not be systematically biased.

Thirdly, experimental controls are lacking. Most research designs require that controls be developed on a *post hoc* basis, and often it is impossible or impractical to provide controls. If the data base is large enough random samples can be taken as controls (e.g., see Jorde and Pitkänen, 1991), but this is a rarity.

Fourthly, there may be temporal and spatial gaps in the data sources that are unavoidable. Such things as accidents (fire, theft, water damage), changes in the personnel who keep vital events records, periods of unrest and warfare, or simply neglect have often resulted in gaps in historical records. These lacunae can be for short periods of time in one set of records or can be spread over a wide area for long periods. In some instances other data sources can be used to supplement and provide missing data. In other instances estimates must be based on data from surrounding time periods or geographical areas. Obviously, any such supplementation needs to be justified and tested.

Size and distribution of the population

The determination of the population size and composition of the group one is studying is of initial importance in many anthropological studies. The size and distribution, and historical changes in these demographic features, can be vitally important for understanding the contemporary genetic structure and disease patterns. Many demographic measures (rates and ratios) are dependent upon accurate estimates of population size and composition. These, in turn, may provide valuable insights into changes in the population and genetic structure over time. Therefore, it is recommended that one obtain as much information concerning the demographic characteristics of the study population as possible.

Basic historical data on population size and composition are usually found in censuses, enumerations, and/or population tables. These sources may only give total size of the population, but may at other times provide detailed listings (see for example Figure 7.2) that divide the population by age, sex, and status (occupation, social status, etc.). These sources need to be checked for accuracy and internal consistency. Once population figures have been checked, there are a number of basic demographic measurements that may be of interest to the researcher. Simple measures fall under the category of rates, ratios, and percentages. Percentages and ratios, such as the population density, sex ratio, and child–women ratio, are useful for analyzing the composition of a population. Rates, on the other hand, are usually used to study the dynamics of temporal change and include such measures as the crude death rate, age–sex-specific death rate, crude rate of natural increase, and standardized birth rate. If one is interested in calculating any of these demographic measures, he/she is strongly advised to consult basic demographic texts before starting since there are conventions that are used by demographers so that data are comparable. The following sources are recommended: Barclay (1958); Henry (1976); Hollingsworth (1969); Palmore and Gardner (1983); Pollard *et al.* (1990); Pressat (1978); Shryock and Siegel (1980); and Willigan and Lynch (1982).

There have been numerous studies of population structure that have effectively used basic demographic data to explore interesting questions that relate to evolution and genetic structure. Population size is an important feature to consider if one is examining founding populations (McCullough and Barton, 1991), rates of gene flow (exogamy and endogamy) and availability of mates (Beckman, 1980; Mielke, 1980; Workman and Jorde, 1980; McCullough, 1989; O'Brien *et al.*, 1989). Populations may be genetically less isolated because of small size, and the mate pools may be very different for small and large populations. Relethford (1986c, 1991) demonstrated that migration distances increase with increasing population size. That is, small populations tend to obtain mates from a relatively restricted mate pool while larger populations have access to more diversified and larger gene pools. Similar, yet slightly different, distributions have been found in other populations (Mielke, 1980; Workman and Jorde, 1980; McCullough, 1989). Relethford (1986a) has also shown that exogamy rates are related to the settlement history and growth of any area. For example, he was able to demonstrate that nonlinear density-dependent migration occurred in historical Massachusetts. Migration was most frequent in small populations (<800) and in large ones (>1600) and the lowest rates were in medium sized groups. Apparently two factors were operating to generate this pattern. For small populations, potential mates were scattered so migration was great; whereas economic factors and transportation seem to explain the increase in migration in large populations. These migration patterns clearly affect such things as inbreeding, genetic drift, the introduction of new alleles into the population, and disease patterns (see also Adams and Smouse, 1985; Relethford, 1985; Adams *et al.*, 1990). Therefore, it is important that basic demographic measures are assembled as an initial step in historical population structure studies.

Mobility and mating structure
For human biologists and physical anthropologists the overriding issues in the study of mating and mobility have been the documentation of the effects of migration on the genetic structure of populations and the determination of population affinities (e.g., see Wijsman and Cavalli-Sforza, 1984). While still of interest, the simple documentation of population (genetic) structure of a specific region has given way to broader interests including such things as examinations of the relationship among demographic parameters (size, distribution, mobility) and the incidence, spread, and impact of various diseases. If one is interested in doing historical research on migration or mobility, he/she should initially define at least three parameters: 1. the space or topography (topology) of the population(s), 2. the movement or nature of mobility, and 3. the characteristics of the migrants.

Assumptions about the space within which the populations are located vary with the models employed and the parameters to be estimated. Most frequently, it is assumed that the space is homogeneous and that migration is isotropic (having the same values in all directions). It is probably generally agreed that the methodology and analysis of mating distances, fission–fusion, and the resulting structure are most appropriately applied to relatively small, regionally subdivided populations. On occasion, nations (Workman *et al.*, 1976) and subcontinents (Menozzi *et al.*, 1978; Ammerman and Cavalli-Sforza, 1984) have been examined. Inference on the genetic structure will require that the probability of migration depends only on distance, that no clines exist, and the population subdivisions are composed of a random collection of individuals. Actual migration rates can be used (Jorde, 1980), thus avoiding the assumption of isotropic movement. Historical data lend themselves well to an examination of the diachronic aspects of settlement patterns in a region and/or the resulting spatial configuration (Cavalli-Sforza, 1962; Küchemann *et al.*, 1967; Swedlund, 1975, 1984; Mielke *et al.*, 1976).

Numerous empirical studies of the structure of human populations have been conducted (for reviews see Jorde, 1980, 1985). As stated earlier, a frequently voiced criticism is that these studies add little more than a supplemental note to local history (Harpending, 1974). One method that can be used to enhance the evolutionary import is to compare and contrast the genetic structures of two or more populations in which specific similarities and differences exist. In a sense, this approach can be viewed as a sort of 'controlled experiment'. Often, differences in structure can be attributed to differences in the demography, culture, ecology, or geography of the populations (Pitkänen *et al.*, 1988, see Box 7.1). Associations, if adequately documented, may lead to useful evolutionary insights. Thus, an examination of the ecological, social, demographic, and political factors (along with the geographical features) that may influence population dispersion, settlement, and affinities is highly recommended (Fix, 1975, 1978; Boyce, 1984; Harding, 1985; Wood *et al.*, 1985; Relethford, 1986b, 1986c; Smith, 1988).

In defining movement, many historical studies have employed migration matrices: matrimonial, parental–offspring, father–offspring, or mother–offspring (Bodmer and Cavalli-Sforza, 1968; Hiorns *et al.*, 1969; Smith, 1969; Morton, 1973; Jorde, 1984). Again, assumptions must be made, such as: that the migration patterns remain constant for a specified period of time (see Wood, 1986 for potential implications), and that migration is essentially Markovian in nature, depending only on the immediately preceding state. Historical studies have an advantage here because the period of time can be broken into discrete sub-units and one can observe temporal trends and changes in the migration patterns and effects on the genetic structure (see e.g., Roberts and Rawling, 1974; Küchmann *et al.*, 1979; Workman and Jorde, 1980; Mielke *et al.*, 1982). The assumption of constant migration rates is affected by differential growth rates of the population subdivisions and changes in mobility brought about by such things as improved transportation networks or economic development in part of the study region. The migration matrix model assumes that migration rates remain constant over time; thus, convergence to equilibrium may take hundreds of generations, and this may not be realistic. Wood (1986) provides an alternative method, which has also shown that migration effectively 'erases' history much more rapidly than had previously been thought. The differences between subpopulations may reflect a few generations of migration and drift rather than deeper historical connections. Also, migration may not be simply Markovian, but influenced by the kinship structure (e.g., see Fix, 1975, 1978 and Kramer, 1981) or other factors. In fact, Fix (1978) has

BOX 7.1.   Marital migration and genetic structure

A frequently voiced criticism of population structure studies has been that
they provide little more than a supplement to the local history of an area
(Harpending, 1974). Therefore, one should attempt to place a study in a
broad context. One way to do this is to compare the genetic structure of two
populations in which specific similarities and differences are known. This
example will detail such an approach using two areas in Finland: 1. the
Åland Island archipelago and 2. the mainland parish of Kitee (see Pitkänen
et al., 1988 for further details and references). Patterns of spatial and
temporal variation in these two populations are assessed and compared
with regard to similarities and differences in ecology, geographic structure,
and social (occupational) structure.

The primary data sources are the Lutheran parish marriage records, poll
tax records, and parish registers. Matrimonial migration matrices were
constructed in order to obtain a measure of genetic differentiation, called
F_{ST} (Wright, 1965; Harpending and Jenkins, 1973).

The F_{ST} values for Åland are in the range for island populations, while
those for Kitee are in the usual range for mainland European populations
(see Jorde, 1980 for a table of comparative values). The F_{ST} values in Åland
are several times greater than those in Kitee for each time period (Table 1).
Also, there is a temporal decrease in Åland's values, but this pattern is not
seen in Kitee.

Some of the reasons for these differences can be explained by examining
differences in endogamy rates, systematic pressures, and effective
population sizes (Table 2). The average of the diagonal elements of the
stochastic migration matrices (p_{ii}, these values reflect endogamy and
indicate the probability that a gene remains in the subdivision of origin) is
much higher in Åland than in Kitee. The systematic pressure values (these
indicate the percentage of individuals originating from outside of the study
area) are several times higher in Kitee than in the Åland Islands. This
would produce comparatively less genetic differentiation in Kitee. Since

Table 1. F_{ST} *values for Åland and Kitee*

Åland		Kitee	
Time period	F_{ST}	Time period	F_{ST}
1750–1799	0.0129	1750–1800	0.0034
1800–1849	0.0115	1801–1840	0.0027
1850–1899	0.0067	1841–1877	0.0028
Total	0.0104	Total	0.0033

Table 2. *Average subdivision population sizes, systematic pressures (migration), and diagonals of the stochastic migration matrices*

Åland

Time period	Size*	Migration*	p_{ii}*
1750–1799	785	0.020	0.910
1800–1849	902	0.018	0.927
1850–1899	1166	0.022	0.928

Kitee

Time period	Size*	Migration*	p_{ii}*
1750–1800	543	0.062	0.762
1801–1840	830	0.048	0.784
1841–1877	1066	0.046	0.834

*These values are averages.

the average size of each subdivision is roughly similar in both populations, genetic drift contributes little to the observed differences in F_{ST} values.

The temporal decline in F_{ST} in the Åland archipelago can also be explained by variables in Table 2. Systematic pressure values increase somewhat in the last period while average p_{ii} values increase slightly over all time periods. The net effect is that internal and external migration contribute little to the genetic variance. At the same time, substantial increases in population size reduce the effects of genetic drift (and, therefore, also genetic variance). In Kitee the forces tend to counterbalance one another – population size increases while both internal and external migration decrease. The result is very little temporal change in Kitee's F_{ST} values.

In summary then, this analysis indicates that: 1. The considerable differences in F_{ST} values between Kitee and Åland reflect greater genetic isolation in virtually all Åland parishes compared with Kitee subdivisions. 2. The differences in F_{ST} values in Åland and Kitee can be attributed primarily to lower migration rates in the Åland archipelago than in Kitee. This reflects greater geographic isolation in the island population than in the mainland one (also, the average geographic distance between population subdivisions in Åland is greater than in Kitee – 34.7 km versus 26.7 km). 3. In addition to geographic differences, Kitee and Åland differ in social organization. Kitee is composed of two church parishes, meaning that people experience social contact with one another through church functions. Åland is composed of 15 distinct parishes, some separated by large expanses of the Baltic Sea, making opportunities for social contact more restricted. Also, villages in eastern Finland were more loosely organized than in western Finland (including Åland). Kitee's villages were rather small, and this appears to have encouraged inter-village marital migration (thus lowering endogamy). On the other hand, a consequence of the tighter village organization and identity in western Finland is reflected in the occurrence of 'village fights' (Haavio-Mannila, 1958), which served to enhance village endogamy.

Thus, by comparing and contrasting two populations in different environmental and cultural settings one gains a perspective that would be lacking in a study of either one alone.

References
Haavio-Mannila, E. (1958). *Kylätappelut. Sosiologinen Tutkimus Suomen Kylätappeluinstituutiosta.* Helsinki-Porvo. WSOY.
Harpending, H. C. (1974). Genetic structure of small populations. *Annual Review of Anthropology*, 3, 229–43.
Harpending, H. C. & Jenkins, T. (1973). !Kung population structure. In *Genetic Distance*, ed. J. F. Crow & C. F. Denniston, pp. 137–61. New York: Plenum Press.
Jorde, L. B. (1980). The genetic structure of subdivided human populations: A review. In *Current Developments in Anthropological Genetics*, Vol. 1, ed. J. H. Mielke & M. H. Crawford, pp. 135–208. New York: Plenum Press.
Pitkänen, K., Jorde, L. B., Mielke, J. H., Fellman, J. O. & Eriksson, A. W. (1988). Marital migration and genetic structure in Kitee, Finland. *Annals Human Biology*, 15, 23–34.
Wright, S. (1965) The interpretation of population structure by F-statistics with special regard to systems of mating. *Evolution*, 19, 395–420.

shown that kin-structured migration may not reduce local (subdivision) variation as would be predicted by most deterministic models of population structure. That is, migration (gene flow) along kinship lines may not reduce gene frequency variances that arise through intergenerational genetic drift. Measuring the effects of non-random migration on population structure may prove to be insightful, and one should examine the statistical method developed by Rogers and Jorde (1987). One should also be aware of the potential genetic effects that can be caused by return migration, i.e., individuals or families moving back to their place of origin (Relethford and Lees, 1983). Thus, a firm understanding of the demographic and social patterns is an important prerequisite for any historical study of mobility/migration and consequent gene flow.

A clear understanding of who is moving is important. It is often assumed that immigration from 'outside' the system is random with respect to genes and that 'within' the system the migrants are randomly drawn from each subdivision. As Dyke (1971) and associates have pointed out, actual mates may deviate substantially from potential mates expected on the basis of local social and demographic structure. This is where historical data can be used to examine deviations from expected patterns. A researcher should attempt to collect data on at least the sex and age distribution, social class or occupations, and fertility and mortality of the migrants versus non-migrants (Küchemann *et al.*, 1974; Hiorns *et al.*, 1977; Cartwright *et al.*, 1978; Hageman *et al.*, 1978; Brennan, 1979; Brennan and Boyce, 1980; Coleman, 1981; Swedlund and Boyce, 1983). Composition factors within subdivided populations remain fertile ground for future investigations because marital migration is seldom random. Also, the genetic effects of movement are still not clearly delineated.

Mate choice, inbreeding, and consanguinity
Demographic, geographic, and sociocultural characteristics of human populations influence and often limit mate availability and selection. Characteristics such as social stratification, male–female ratio, age structure, size of the population, geographic distribution, and migration patterns can impose constraints on mate selection (see e.g., Epstein and Guttman, 1984). If one is interested in studying the influence of these features on the patterns of mating, potential mates analysis (PMA) may be extremely useful. Temporal changes in these influences can be documented by incorporating historical data into the analysis.

Potential mates analysis is a technique that examines the effects and importance of various population characteristics (sociocultural, demographic, and geographic) that influence mate availability and choice, and which, in turn, influence population (genetic) structure (Dyke, 1971; Leslie, 1985; O'Brien *et al.*, 1989; Williams-Blangero, 1990). The technique involves comparing the characteristics of actual mates with those of potential mates by establishing a potential mate pool for each individual. In doing so, mating rules can often be inferred (Dyke, 1971). Simulation and actual mate pools may be incorporated into the analysis if data are available (Frankenburg, 1990). Demographic characteristics such as age at marriage and age differences between spouses may document changes in composition of the population which in turn may influence mate choice.

Williams-Blangero (1990), using PMA, was able to examine the influences of demographic (male–female age composition) and kin structures (rules of exogamy and endogamy) on mate choice among the Jirels of eastern Nepal. O'Brien *et al.* (1989) have shown how genealogies, kinship analysis, and PMA were used to understand the temporal changes in mate pools, mate selection, and consanguinity avoidance from 1700 to 1950 in Sottunga, Finland. No significant differences or consistent temporal trends in mate pools were found between occupational groupings of farmers and non-farmers, even though farmers were slightly more related to their mates and more related among themselves than nonfarmers. Potential mates analysis can detect causes of nonrandom mating such as population size and age structure constraints (Leslie, 1985) and can document preferential cousin marriage and the influence of neighborhood knowledge (Frankenburg, 1990). Frankenburg was able to identify the actual mate pool in a historical sample and compare it to the actual mating patterns in a community in Virginia during the period from 1850 to 1939. She was able to document and examine temporal trends in nonrandoming mating patterns such as positive assortative mating for cousins, nuclear kin avoidance, and age difference preferences for mates.

PMA analysis can often be supplemented and extended in interesting ways. For example, the type of analysis conducted by James and Morrill (1990) on culturally irregular mating choices could be effectively added to a PMA analysis. In a study of St. Bart in the French West Indies, these authors examined different types of matings that violated indigenous cultural rules (illegitimate children, first cousin marriages) or were deviations from statistical norms (e.g., large age differences in spouses, late age at marriage). Interestingly, these mating irregularities were found to run in families.

Isonymy

One of the most convenient methods for estimating inbreeding is also a method that can use information that is readily available in historical sources – that is, inbreeding coefficients estimated from marital isonymy. Isonymy refers, literally, to 'same name'. The original work on this procedure (Crow and Mange, 1965) took into account the fact that people sharing the same surname are potentially related and showed how the proportion of isonymous marriages in a population could suggest the degree of inbreeding in that population. For a more complete discussion of these methods and their application the reader is referred to Crow (1980, 1983); Gottlieb (1983); Lasker (1985, 1988); Lasker *et al.* (1986); Jorde and Morgan (1987); Relethford (1988); Jorde (1989); and Kosten and Mitchell (1990).

It can be shown that under certain assumptions the relationship of inbreeding (F) to isonymy (I) is:

$$F = \frac{I}{4}$$

'I' is defined here as the proportion of same-name marriages in the population.

Inbreeding, estimated by isonymy, is normally assumed to be composed of a random component (F_r) that occurs simply as a function of the numbers of individuals and proportions of each sex in the population:

$$F_r = \Sigma p_i q_i / 4$$

where p_i is the frequency of the ith surname in fathers and q_i is the frequency of the ith surname of mothers (maiden name). In addition, it is possible to estimate a nonrandom component, which may be the result of intentional preference or avoidance. The nonrandom component (F_n) is estimated by:

$$F_n = (I - \Sigma p_i q_i)/[4(1 - \Sigma p_i q_i)]$$

By careful analysis of actual marriage records in a population it is possible to refine considerably the empirical estimates of inbreeding in a population and also to understand better the processes by which inbreeding comes about. Also, if growth rates and other aspects of population structure are understood, then one can avoid some of the pitfalls of estimating inbreeding using only the theoretical relationships. Considerable controversy still persists about how accurate isonymy is for estimating inbreeding, but several studies in the past few years have suggested that it can be a valid estimate, particularly when accompanied with other information regarding the demography of the community(ies) in question (see Gottlieb, 1983; Jorde, 1989). However, see Rogers (1991) for some important implications if certain assumptions are not met.

Another aspect of surname analysis is that surnames, unlike genes, have phenotypic and cognitive properties apparent to the people who are actually mating. This allows one to interpret some of the social and geographic processes that have contributed to observed isonymy that might be obscured when using strictly genetic inferences.

In the Connecticut Valley of Massachusetts surnames have been used to look at a host of questions regarding historical population structure (Swedlund and Boyce, 1983; Swedlund, 1984; Swedlund et al., 1984). Because additional data were available on who the individuals were who possessed these surnames, and because their relative economic status in the population was known, it was possible to move through a series of questions that included:

1. Basic estimates of inbreeding for a series of communities.
2. Analysis of the geographic distribution of surnames in these communities and the regional levels of inbreeding.
3. Estimates of the likelihood that members of different communities would mate.
4. Analyses of intermarriage rates within the elite families of the region.

The study of isonymy is an attractive option for those pursuing historical studies. However, one must be cautious in interpreting results, and realize that for the surname estimates to be truly reflective of genetic processes the names must not have multiple origins, the founding stock must be specified, and the populations must be reasonably stable (Crow, 1980, 1983; Rogers, 1991). To the extent that basic assumptions are met, however, isonymy affords a simple and straightforward approach to one estimation of inbreeding, at least in relative terms.

Natural selection in historical populations
If a population has:

(a)   *variability* (trait or characteristic diversity among individuals),
(b)   *fitness differences* (a relatively invariable association between a trait or attribute and fertility, fecundity, and/or survivorship), and
(c)   *inheritance* of the trait or characteristic that is, at least, partially independent of environmental influences, then, as stated by Endler (1986, p. 4):

> 1. the trait frequency distribution will differ among age classes or life-history stages, beyond that expected from ontogeny;
> 2. if the population is not at equilibrium, then the trait distribution of all offspring in the population will be predictably different from that of all parents, beyond that expected from conditions a and c alone.

Thus, selection is of obvious demographic interest. Detecting and measuring natural selection in human populations is, however, very difficult (Bajema, 1963; Bodmer, 1968; Endler, 1986; Jorde and Durbize, 1986) and requires extremely large sample sizes (Barrett, 1990). Endler (1986) presents ten methods for detecting natural selection in naturally occurring populations. Most, if not all, of these methods use genetic data, making the procedures of limited use to the historical demographer interested in demonstrating the possible action of natural selection. There are, however, at least four approaches that have been developed that utilize historical demographic data:

1. Direct measure of fitness
2. Intrinsic rate of natural increase
3. Crow's Index
4. Cohort method.

Direct measure of fitness

This method allows one to calculate the relative reproductive fitness of a particular phenotype by examining the ratio of the average number of offspring produced by affected persons to the average number of offspring produced by 'normal' individuals. The most widely cited application of this technique is Morch's (1941) research on achondroplasia. The relative reproductive fitness of the individuals in this sample was 0.1963 or about 20% (calculated by Popham, 1953). Later, Cotter (1967) examined the hypothesis that the relative fitness of the dominant gene producing achondroplasia is also a function of culture. By using pedigrees from Utah Mormons in the nineteenth century, he obtained a fitness value of 0.622. Cotter then argued that one should only compare those individuals who actually contribute to the next generation (keep in mind, however, that *fitness,* by definition, includes individuals who do not mate and reproduce successfully). After this correction, he found that the average number of offspring per anchondroplastic was not significantly different from 'normal' individuals and thus concluded that relative fitness in this case was a function of culture–the fitness value varies as the attitudes of the breeding population change over time (or from culture to culture). The major problem with this approach for the historical demographer is that it requires large samples and that 'normal' and 'affected' individuals are easily and accurately identified in records, and as such it is usually limited to identifiable phenotypic differences.

Intrinsic rate of natural increase (IRNI)

In 1930 Fisher suggested combining fertility and mortality rates of a population with factorial inheritance so that the principle of natural selection could be expressed in a precise mathematical theorem. The difference between fertility and mortality rates regulates increases or decreases in a population and can be used to assess the rate of 'improvement' of a population in relation to its environment. This method then employs the demographic measure called the intrinsic rate of increase (IRNI). This rate is calculated for both the 'affected' and 'normal' portions of the population, and then the rates are compared to determine fitness of the two groups:

$$W_i = \frac{e^{r_{mi}T}}{e^{r_{mh}T}}$$

where e is the base of the natural logarithm, T is the average generation length for the total sample, r_{mi} is the IRNI for subgroup i, and r_{mh} is the IRNI for subgroup having largest r_m (IRNA) value.

Bajema (1963) was the first to apply IRNI to estimate the direction and intensity of natural selection on IQ, finding that there was a bimodal relationship between fertility and IQ in a sample from Kalamazoo, Michigan. He strongly cautioned that one should not generalize from these results. The method was also applied by Bodmer (1968) who found that male and female schizophrenics were at a selective disadvantage. Even though Bajema (1963), Bodmer (1968) and Cavalli-Sforza and Bodmer (1971) have recommended the use of this method, few studies have actually been done (Adams and Smouse, 1985; Roberts, 1988). The method is of interest to historical demographers because it takes into account fertility, mortality, and generation length. The major problem with this technique is that it is again difficult to detect different genotypes in historical data and then trace those individuals through their entire life course.

Crow's Index
With few exceptions the study of natural selection in historical populations has been done without regard to tracing the fertility and mortality profiles of specific individuals, but rather, analysis has consisted of collecting aggregate profiles of mean variances in fertility and survivorship. The calculation of Crow's Index is the method most frequently employed to examine *potential* for natural selection in historical populations (Crow, 1958, 1989). This index stems from Fisher's (1930, p. 35) fundamental theorem of natural selection, which states that, 'the rate of increase in fitness of any organism at any time is equal to its genetic variance at that time'. Crow's Index sets an *upper limit* on the amount of change in gene frequency that selection can cause. The index is calculated by dividing the variance in number of offspring by the square of mean number of offspring. The index is usually divided into fertility (I_f) and mortality (I_m) components:

$$I = I_m + (1/P_s)I_f$$

where

$$I_m = P_d/P_s$$

and

$$I_f = V_f/\bar{X}_s^2$$

where P_d = (prereproductive mortality) the proportion of 'would be' parents who die before age 15, P_s = proportion who survive beyond age 15, V_f = variance in fertility, and $\bar{X}_s^2$ = square of the mean number of offspring born to the proportion of 'would be' parents who survive.

Or the mortality component:

$$I_m = \frac{1-S}{S}$$

where S = the proportion of individuals who survive to age 15.
Then,

$$I_t = I_m + I_f/S$$

A number of modifications of Crow's Index (original formula) have been suggested to alleviate some of the simplifying assumptions (e.g., Kobayashi, 1969; Crow, 1973; Spuhler, 1976):

1.   Maternal mortality during the childbearing period may be added.
2.   Prenatal and postnatal mortality can be separated.
3.   Embryonic mortality may be added.
4.   Different distributions can be used to compute means and variances.

Instead of using completed family size, 'successful' offspring can be used to calculate the fertility component. Success can then be defined in different ways: 1. those who reach 15, or 2. those who marry, or 3. those who contribute to the next generation. The researcher should also be cognizant of the fact that much of the variation in who marries and/or reproduces (and how many children) is very likely to be influenced by cultural factors.

Use of the index has been reviewed by Spuhler (1976) and Ward and Weiss (1976). An exhaustive list of values for different populations around the world is also provided in Spuhler. While this method affords the investigator with a means of estimating selection potential, much of the variance in fertility and mortality can and probably does result from cultural and non-genetic environmental factors. Nevertheless, it does provide some inference on selection when genotypic/phenotypic data are lacking. Usually single estimates from one population at one point in time are meaningless. However, when cross-cultural comparisons or temporal trends in the indices are observed, the parameters may provide some insights (Morgan, 1973; Crawford and Goldstein, 1975; Tempkin-Greener and Swedlund, 1978; Hed, 1984, 1986a, 1986b, 1987; Hed and Rasmuson, 1981; Koertvelyessy, 1983; Adams and Smouse, 1985; Dinsmore, 1985; Jorde and Durbize, 1986; Reddy and Chopra, 1990). Modernization usually tends to lower the index, both through lower mortality rates and by a reduction in fertility (Jacquard and Ward, 1976; Spuhler, 1976; Hed, 1987). Fertility has become increasingly important in the opportunity for selection as medicine and public health innovations have reduced infant mortality substantially.

Cohort method

A fourth approach that has seen limited application involves the use of cohort data to measure differences in survivorship in two comparable samples from the same population (see Box 7.2). This approach capitalizes on historical events that afford the opportunity to estimate fitness differences in phenotypes, as measured by response variation to disease. One difficulty with the cohort method is that it requires known birthdate and age at death of each individual to be assessed. Moreover, it requires that each individual be a member of a cohort that was exposed to an infectious disease or common insult. It is also best applied when one can obtain information on a cohort that otherwise resembles the exposed group (cohort) but did not experience exposure to the disease.

In some respects the cohort method loosely resembles a hybrid of the 'cohort' and 'case controlled' study designs in epidemiology (e.g. Kleinbaum et al., 1982). The difficulties in obtaining the requisite data are outweighed by the greater precision with which one can assess differences in survivorship. Whereas Crow's Index only measures the potential for selection to exist, the cohort method does estimate actual differences in survivorship resulting from exposure to disease.

The ideal experiment is one in which a cohort is exposed to an epidemic at a particular point in time (see Meindl and Swedlund, 1977). Those who do not survive the epidemic are removed from the analysis and a life table is constructed for all those who died at some predetermined point after the epidemic and presumably independently of exposure to the epidemic. This 'exposed' cohort is then compared to a 'control' cohort which

BOX 7.2. Stressed cohorts and fitness

One analysis of fitness, in its broadest sense, that has been undertaken with historical data is the analysis of stressed cohorts in the Connecticut Valley of Massachusetts. We wished to find a precise way to measure the impact of a diarrheal epidemic on an exposed cohort relative to a cohort that did not experience the episode but that in other ways was comparable (Meindl and Swedlund, 1977). Using data from an 1802 epidemic in Greenfield and an 1803 epidemic in Deerfield we observed the casualties and then asked the question, what is the mortality experience of the survivors of these epidemics in relation to their 'unstressed' counterparts? The control groups in this case were comprised of children born after the epidemic in question and at a time in which they were not exposed to an epidemic during early childhood.

Our question revolved around whether the survivors of an epidemic might not be somehow more constitutionally fit than their stricken cohort members, and perhaps even more fit on average than the control group that had not been stricken.

After controlling for age and secular trends we did indeed find that the survivors of epidemics experienced, overall, longer average life expectancy than the controls. Their 'constitution', whether it was genetic or acquired, seemed to predispose them to relatively greater longevity, and this was true for both communities. Both males and females showed the relative increased survivorship among the stressed cohort members. A second part of this question had to do with whether or not the exposure to the epidemic affected their basic genetic-immune system and if these changes were conferred to their offspring resulting in increased fertility or longevity evidenced in their children and grandchildren. No evidence could be detected for such an outcome, though small sample sizes were definitely a problem (Meindl, 1984). Nor should one necessarily expect this to occur given the fact that the non-specific nature of the epidemics themselves was not likely to screen the genotypes effectively and general immune competence might well have related more to adequate nutrition and good health in the survivors, as alluded to above.

Nevertheless, we believe that the stressed cohort model is an effective means for approaching a variety of questions about differential survivorship in many populations. It has been used to some advantage on an historical population in Finland (Trapp, 1981; Mielke *et al.,* 1987) and offers a sound methodological approach regardless of whether the hypothesis about the stress is one of conferring an advantage or one of only conferring another insult on the hosts in question. Historical epidemiological studies need to make better use of the retrospective research designs used in contemporary epidemiological research.

(continued)

References
Meindl, R. S. (1984) Components of longevity: Developmental and genetic responses to differential childhood mortality. *Social Science and Medicine*, **16**, 165–74.
Meindl, R. S. & Swedlund, A. S. (1977) Secular trends in the Connecticut Valley, 1700–1850. *Human Biology*, **49**, 389–414.
Mielke, J. H., Pitkänen, K. J., Jorde, L. B., Fellman, J. O. & Eriksson, A. W. (1987). Demographic patterns in the Åland Islands, Finland, 1750–1900. *Yearbook of Population Research in Finland*, **25**, 57–74.
Trapp, P. G. (1981). The Demographic Impact of Smallpox Mortality on the Population of the Åland Islands, Finland 1750–1890. MA Thesis, Dept. of Anthropology, University of Kansas.

resembles the exposed cohort very closely in time and space, demography, and environment, but which was not exposed to the epidemic (e.g., a group of individuals born immediately after the epidemic struck the community). It is also necessary to estimate survivorship for each group commencing at an age that is *older* than the age at which the individuals were when exposed. For example, if measuring the impact of a childhood epidemic that struck a cohort of children when they were 3–5 years of age, then *subsequent* survivorship of those who lived would be estimated from age 10 years onwards. The control cohort's survivorship would also be estimated for ages 10 and older. By comparing the survivorship to each subsequent age and measuring the difference, an estimate of the effects of the epidemic on cohort longevity is obtained.

One then assumes that the summed differences in survivorship, or the mean longevity in years, are measures of difference in the two cohorts resulting from the epidemic exposure. If the 'exposed' cohort survivors indicate greater longevity than their 'controls', then one might assume some enhanced fitness or other constitutional advantage enabled them to survive the epidemic and that the 'weaker' or less fit individuals of the 'exposed' cohort died as a result of exposure. Conversely, if the 'control' cohort showed greater mean survivorship or mean longevity, then no fitness effects can be assumed. If the 'control' cohort indicated greater survivorship or mean longevity than the 'exposed', then one might assume or look for evidence that the epidemic *weakened* the survivors as well as eliminating the non-survivors. The real differences are generally assessed by means of non-parametric tests such as the Kolmogorv–Smirnov test (eg. Lee, 1980).

The cohort approach is reviewed in Meindl and Swedlund (1977), Meindl (1984), and Goodman *et al.* (1988) for the Connecticut Valley and in Mielke *et al.* (1987) for Finland. Many refinements and alternative designs using elements of the cohort method are possible in those cases where good historical data are available. The method is attractive in those cases where a genetic susceptibility to an attributed disease is known, or where immune response to a particular disease can be assumed to be genetically influenced. However, in historical studies these assumptions are often difficult.

In summary, Crow's Index, the intrinsic growth rate, the cohort, and other methods provide inference on the potential for selection in historical populations. Their utility is probably maximized when temporal trends are considered rather than relying upon a single estimate. At best, an overall picture of the contributions by mortality and/or fertility may be achieved, something that demographic trends might also provide but perhaps with less resolution. In the case of the cohort method some additional insight may be obtained regarding differential mortality in response to a specific selective environment.

Historical epidemiology

Infectious (epidemic) diseases have often played an important role not only in determining the mortality patterns in a population but also in shaping the demographic structure. A variety of deterministic and stochastic models have been developed to examine the population dynamics of disease spread (Bartlett, 1956; Bailey, 1975; Anderson and May, 1979; Anderson, 1982). These models include such factors as probability of contact, infectivity, latency, and other features that relate to general dynamics of subdivided populations. Much can be learned about the characteristics of epidemics by studying them in historical, well-defined populations. Black (1966), Cliff *et al.* (1981), and Cliff and Haggett (1984) have shown that the duration and periodicity of measles epidemics are affected by both population size and density. Other studies have focused on the influence and importance of medical intervention in the declining mortality rates over the last few centuries (Preston, 1976; McKeown, 1976, 1979; Cherry, 1980; Collins, 1982). In recent years there has been increasing interest in anthropology in applying insights gained from population structure models to questions of disease impact and geographic spread.

Much work has been done on the geographic spread of disease in prehistoric, historic, and contemporary contexts. Some papers have been primarily speculative or suggestive such as those elucidating the changing disease patterns in prehistory (Cockburn, 1967, 1971; Armelagos and Dewey, 1970; Armelagos and McArdle, 1975) or hypothesizing about the spread of such diseases as syphilis from one continent to another (Hudson, 1965; Crosby, 1972; Baker and Armelagos, 1988). Documenting the changing disease patterns that accompanied the transition from hunting/ gathering to food production has attracted much attention recently (Cohen and Armelagos, 1984; Swedlund and Armelagos, 1990). Some modeling has been applied to the question of the impact of disease on Native American populations (Upham, 1986; Ramenofsky, 1987). McGrath (1988) has shown that effective population size was an important factor in the endemicity of tuberculosis in the Lower Illinois valley. On the other hand, the social structure, especially the degree of contact between peoples, was the most important factor in disease spread. Recently, McGrath (1991) focused her attention on the biological impact of the social disruption that accompanies many epidemic diseases. She proposed a model of varied responses that reflects the effects of increasing social disruption on the biological impact of disease. The prospect of using this model in historical studies is intriguing and well worth developing.

Epidemiologic studies have often focused on modeling disease spread or predicting the diffusion of epidemics within populations (Bartlett, 1956; Bailey, 1975; Cliff *et al.*, 1981; Sattenspiel, 1987a). Often the data required for such models (e.g., information on the number of susceptible individuals in the population) do not exist in historical data sets and alternative strategies must be sought. Such studies have, however, clearly shown that population structure and social interaction are critical determinants in the spread of disease within and among populations (e.g., see Cliff *et al.*, 1981; Cliff and Haggett, 1984; Sattenspiel, 1987b, 1988, 1990; Sattenspiel *et al.*, 1990).

Epidemiologic studies are often confined to rather short periods of time. There is, however, much to be learned from studying the long term temporal behavior of diseases in well-defined populations. There has been research along these lines directed to the question of the effect that infectious diseases have had on the demographic structure of populations. Using historical data, Mercer (1985) has examined the demographic effect of smallpox vaccination in European populations. May *et al.* (1988, 1989) have looked at the impact AIDS may have on the demographic structure of a population. Research on historical epidemiology in Finland has also shed some light on the demographic impact that epidemics have on population structure (see Box 7.3 and Mielke *et al.*, 1984; Jorde *et al.*, 1989, 1990; Pitkänen *et al.*, 1989).

Famine induced crisis mortality may also have an impact on the demographic structure of a region that, in turn, may influence the genetic structure. Bittles *et al.* (1986) and Bittles (1988), using censal data, have shown that the Irish Famine (1846–1851) was responsible for creating

BOX 7.3.   Historical epidemiology of smallpox in the Åland Islands

The relative importance and role of medical advances, including immuniz-ation, in the declining mortality of the last two centuries is a source of continuing controversy. As part of this historical problem, we present an analysis of one disease (smallpox) that appears to have responded to early medical intervention (see Mielke *et al.*, 1984 for more details and refer-ences). Similar analyses could be conducted with other disease entities provided one has access to longitudinal, cause-specific mortality data.

Our focus in this example is the Åland Island archipelago that lies between Sweden and Finland in the Baltic Sea. Smallpox mortality data (date, age, sex, and place of death) were collected from each of Åland's 15 Lutheran parishes for the period 1750–1900. We also collected aggregate level statistics such as total number of deaths per year to use for comparison and annual number of children vaccinated from 1805 onwards (used to estimate the proportion of children vaccinated per year).

By plotting the data each year it was shown that there were 19 epidemics that affected the archipelago. Proportionate mortality ratios (PMR) demonstrated that from 0.70% to 30.05% of the deaths during epidemic periods could be attributed to smallpox. There was also a decrease in the PMRs after 1800. This decrease was accompanied by a shift in the age pattern of mortality with a reduction in the proportion of deaths among children aged one to 15 (72.9% between 1750–1800 and 26.7% after 1800). This characteristic is precisely the type of change one would expect because most children were vaccinated during their first year of life and the effect lasted about ten years. Also, a larger proportion of deaths occurred in the adult population, and for the first time adults over 60 years of age were affected. We can suggest two reasons for this occurrence. Firstly, vacci-nation could have reduced the general exposure of the population to such a degree that a sufficient number of susceptible adults was finally generated (i.e., individuals neither vaccinated nor exposed to the virus as children). Secondly, vaccination did not provide lifetime immunity. If revaccination of adults was not common, an adult population vulnerable to smallpox infection would have existed. Both of these factors appear to have operated in Åland.

Visual inspection of the time series of smallpox mortality (i.e., the number of deaths plotted per year) indicated a regular temporal pattern (see the figure). This impression was confirmed statistically by spectral analysis, which demonstrated that there was a strong seven year period-icity. This regular cycle probably represents the amount of time required to accumulate a cohort of susceptible individuals that is sufficiently large to allow the smallpox virus to spread effectively over the archipelago. The changes in PMRs and the age shift suggested that a separate examination of

(continued)

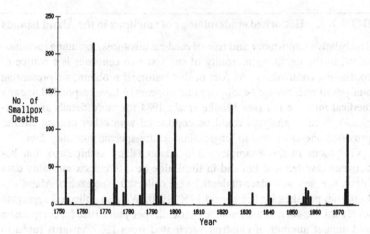

Number of smallpox deaths per year in Åland from 1750 to 1775.

the periodicity for the period 1750–1800 (pre-vaccination) and 1800–1900 (post-vaccination) was warranted. As expected, the pre-vaccination spectral density plot showed a strong peak at seven years. The post-vaccination plot revealed a strong periodicity in the same region, but it was at eight years rather than seven. The lengthening of the dominant periodicity may reflect the influence of smallpox vaccination, which extended the amount of time needed to build up a susceptible cohort. The most important observation is that there was still periodic behavior in the post-1800 series, although because of vaccination the amplitude (i.e., variance) of this series was markedly reduced. Vaccination appears to reduce the severity of the outbreaks, but it does not eliminate the periodicity.

Earlier analyses of the population structure of the Åland Islands utilizing matrimonial migration matrices and genetic data demonstrated that there was both historical and contemporary genetic heterogeneity among the 15 parishes. This heterogeneity implies that there could be regional variation in the impact of smallpox. Marital exchange data suggested that there may be parishes that, because of their isolation, were not affected by each epidemic. To test this assumption, a measure of isolation was constructed by multiplying the population size of each parish by the proportion of each parish population that consisted of migrants. To measure regional variation in epidemic events, the number of epidemics experienced by a given parish was divided by the total number of epidemics that took place in Åland. A Spearman's rank-order correlation of 0.918 ($p < 0.001$) between these two measures indicated a good correspondence between epidemics and general isolation (population structure) of the subdivisions.

This example shows how historical demographic data, genetics, and epidemiology can be combined to provide insights into historical and contemporary problems of epidemic disease spread and control. These types of analyses could be used to study the dynamics of other infectious diseases under different ecological, spatial, and temporal conditions.

Reference
Mielke, J. H., Jorde, L. B., Trapp, P. G., Anderton, D. L., Pitkänen, K. & Eriksson, A. W. (1984) Historical epidemiology of smallpox in Åland, Finland: 1751–1890. *Demography*, **21**, 271–95.

marked differences in population structure among subdivisions and regions. Effects on household size, population density, effective population size, and male/female ratios were not consistent from region to region despite massive population losses. They suggest that high rates of recessive genetic disorders in the contemporary population can be partially attributed to these famine-induced changes. Smith *et al.* (1990), using isonymic analysis, have also documented the potential genetic consequences of the Irish Famine on population subdivisions in North Eastern Ireland. They found that there were discrete breeding groups with minimal genetic exchange. Coupled with political–religious and geographic barriers to gene flow, this post-famine structure enhanced the potential for genetic drift and inbreeding.

Clearly, a long term perspective on population change may be essential for understanding the present day distribution of diseases.

Summary and conclusions

Historically based studies have made important contributions to our understanding of the biology of human populations. Assumptions and predictions regarding the extent of inbreeding in traditional human populations have been addressed, and the range of possible values are now well understood, largely as a result of studies in historical population structure.

So far, the effects of mutation and selection have been less well measured in historical studies. As described in this chapter, the methods developed thus far has been based on differential mortality and fertility, which, while provocative, cannot measure or estimate real genotypic or phenotypic differences. Perhaps future research will, if based on deep and accurate genealogical data, provide insights. We should not be too optimistic however, for while we have some intriguing data and problems to explore, most of these revolve around questions of the coevolution of disease pathogens and human hosts. The human genetics of these interactions generally involve the immune system, where there is still so much to be learned.

Svanborg-Eden and Levin (1990) have recently argued that in the coevolution of pathogen and host it is generally the pathogen that will exhibit genetic heterogeneity and evolutionary change because of the very rapid generation times of microbial bacteria and viruses. It is difficult for the human genome to respond in specific and effective ways to the ever-changing pathogen, given our generation spans. Moreover, Barrett (1990) has demonstrated that even if human adaptation to a specific pathogen does occur, the ability to measure that genetic response requires sufficiently large sample sizes to detect significant differences in survival rate or susceptibility as to rule out historical approaches and even most research contexts involving living populations. Therefore, we must conclude that opportunities for measuring selection in historical populations are all but ruled out by the absence of necessary conditions and data. Disease mediated selection is just too problematic.

One small but tantalizing exception to this limitation might be to find the kind of genetic polymorphism that confers an advantage to some genotypes over others. Such is the case with sickle cell disease (and possibly the Duffy antigen system) in the presence of malaria. This example is well understood at the level of the gene, the population, and the disease process. Such cases are not presumably the result of disease mediated selection but rather result from previously existing genetic conditions that have a fortuitous effect in the presence of a disease. Good demographic and genealogical data might then permit some inference.

It has been suggested that such an example might exist with the genetic disease cystic fibrosis (CF) and tuberculosis (e.g., Crawfurd, 1972; Meindl, 1987), as well as other diseases. There is biochemical and clinical evidence that heterozygote carriers of the cystic fibrosis gene might have a selective advantage in the presence of the tubercle bacillus, and that this might in turn explain the relatively high levels of cystic fibrosis historically in areas of north and western Europe where tuberculosis was rampant in the past. Jorde and Lathrop (1988) were unable to detect any fertility advantage in CF carriers and thus suggest that random genetic drift or a past selective event (e.g., lower mortality rate in CF carriers) are more likely to be responsible for the observed CF gene frequencies. Nonetheless, hypothesized associations such as these are intriguing and worthy of study using historical data.

Whether or not selection–mutation processes will be better understood through studies of historical populations is not, however, the major issue. The lessons we have learned from historical approaches are many and have considerably enriched our appreciation of human population processes. The study of historical population structure and of historical epidemiology will assuredly provide new insights into the history of disease and microevolution for many years to come. As new methodologies are developed and additional populations are analyzed, the mosaic of human biology in the past will become increasingly vivid.

Acknowledgements
We thank Drs. Lynn Jorde, Gabriel Lasker, and Nicholas Mascie-Taylor for their constructive comments and suggestions. This work was supported in part by NSF grant BNS-8319057 (JHM); the Sigrid Jusélius Foundation, Helsinki, Finland; and Samfundet Folkhälsans Genetiska Institut, Populationsgenetiska Avdelningen, Helsinki, Finland.

References
Adams, J. W. & Kasakoff, A. B. (1980). Migration and marriage in colonial New England: A comparison of rates derived from genealogies and rates from vital records. In *Genealogical Demography*, ed. B. Dyke & W. T. Morrill, pp. 115–38. New York: Academic Press.

Adams, J. & Smouse, P. E. (1985). Genetic consequences of demographic change in human populations. In *Diseases of Complex Etiology in Small Populations: Ethnic Differences and Research Approaches*, ed. R. Chakraborty & E. J. E. Szathmary, pp. 283–99. New York: Alan R. Liss.

Adams, J., Lam, D. A., Hermalin, A. I. & Smouse, P. (eds.) (1990). *Convergent Issues in Genetics and Demography*. New York: Oxford University Press.

Åkerman, S., Johansen, H. C. & Gaunt, D. (eds.) (1978). *Chance and Change*. Odense: Odense University Press.

Alström, C. H. & Lindelius, R. (1966). *A Study of the Population Movement in Nine Swedish Subpopulations in 1800–1849 from the Genetic-statistical Viewpoint*. Basel: S. Karger.

Ammerman, A. J. & Cavalli-Sforza, L. L. (1984). *The Neolithic Transition and the Genetics of Populations in Europe*. Princeton: Princeton University Press.

Anderson, R. M. (1982). *The Population Dynamics of Infectious Diseases: Theory and Application*. London: Chapman and Hall.

Anderson, R. M. & May, R. M. (1979). Population biology of infectious diseases. *Nature*, **280**, 361.

Armelagos, G. J. & Dewey, J. (1970). Evolutionary response to human infectious disease. *Bioscience*, **20**, 271–5.

Armelagos, G. J. & McArdle, A. (1975). Population, disease, and evolution. *American Antiquity*, **40**(2) (Memoir 30), 1–10.

Bailey, N. T. J. (1975). *The Mathematical Theory of Infectious Diseases*. London: Griffin.

Bajema, C. (1963). Estimation of the direction and intensity of natural selection in relation to human intelligence by means of the intrinsic rate of natural increase. *Eugenics Quarterly*, **10**, 175–87.

Baker, B. J. & Armelagos, G. J. (1988). The origin and antiquity of syphilis. *Current Anthropology*, **29**, 703–37.

Baker, P. T. & Sanders, W. T. (1972) Demographic studies in anthropology. *Annual Review of Anthropology*, 1, 151–78.

Barclay, G. W. (1958). *Techniques of Population Analysis*. New York: John Wiley & Sons.

Barrett, J. A. (1990). The detection of selective differences in populations. In *Disease in Populations in Transition*, ed. A. C. Swedlund & G. J. Armelagos, pp. 47–53. New York: Bergin & Garvey.

Bartlett, M. S. (1956). Deterministic and stochastic models for recurrent epidemics. *Proceeding of the Third Berkeley Symposium on Mathematical Statistics and Probability*, 4, 81–100.

Beckman, L. (1980). Time trends in endogamy rates in northern Sweden. In *Population Structure and Genetic Disorders*, ed. A. W. Eriksson, H. Forsius, H. R. Nevanlinna, P. L. Workman & R. K. Norio, pp. 73–80. London: Academic Press.

Beckman, L. & Cedergren, B. (1971). Population studies in Northern Sweden. I. Variation of matrimonial migration distances in time and space. *Hereditas*, 68, 137–42.

Bittles, A. H. (1988). Famine and man: Demographic and genetic effects of the Irish famine, 1846–1851. In *Anthropologie et Histoire ou Anthropologie Historique?* ed. Luc Buchet, pp. 159–75. Actes des Troisiémes Journées Anthropologiques de Valbonne, Notes et monographies techniques No. 24. Paris: Centre National de la Recherche Scientifique.

Bittles, A. H., McHugh, J. J. & Makov, E. (1986). The Irish Famine and its sequel: population structure changes in the Ards Peninsula, Co. Down, 1841–1911. *Annals Hum. Biol.*, 13, 473–87.

Black, F. L. (1966). Measles endemicity in insular populations: Critical community size and its evolutionary implication. *J. Theor. Biol.*, 11, 207–11.

Bocquet-Appel, J-P. & Jakobi, L. (1990). Familial transmission of longevity. *Annals of Human Biology*, 17, 81–95.

Bodmer, W. F. (1968). Demographic approaches to the measurement of differential selection in human populations. *Proceedings of the National Academy of Sciences*, 59, 690–9.

Bodmer, W. F. & Cavalli-Sforza, L. L. (1968). A migration matrix model for the study of random genetic drift. *Genetics*, 59, 565–92.

Boyce, A. J. (1984). *Migration and Mobility: Biosocial Aspects of Human Movement*. London: Taylor & Francis.

Brennan, E. R. (1979). Kinship, demographic, social and geographic characteristics of mate choice in a small population. PhD Dissertation, Pennsylvania State University, State College, PA.

Brennan, E. R. & Boyce, A. J. (1980). Mate choice and marriage in Sanday, Orkney Islands. In *Genealogical Demography*, ed. B. Dyke & W. T. Morrill, pp. 197–207. New York: Academic Press.

Caldwell, J., Caldwell, P. & Caldwell, B. (1987). Anthropology and demography. *Current Anthropology*, 28, 25–43.

Cannings, C. & Cavalli-Sforza, L. L. (1973). Human population structure. *Advances in Human Genetics*, 4, 105–72.

Carmelli, D. & Cavalli-Sforza, L. L. (1976). Some models of population structure and evolution. *Theoretical Population Biology*, 9, 329–59.

Carr-Saunders, A. M. (1922). *The Population Problem: A Study in Human Evolution.* London: Oxford University Press.

Cartwright, R. A., Hargreaves, H. J. & Sunderland, E. (1978). Social identity and genetic variability. *Journal of Biosocial Science,* 10, 23–33.

Cavalli-Sforza, L. L. (1959). Some data on the genetic structure of human populations. *Proceedings of the X International Congress in Genetics,* 1, 389–407.

Cavalli-Sforza, L. L. (1962). The distribution of migration distances; models and applications. In *Les Desplacements Humains,* ed. J. Sutter, pp. 139–58. Paris: Hachette.

Cavalli-Sforza, (1969). 'Genetic drift' in an Italian population. *Scientific American,* 221, 30–7.

Cavalli, Sforza, L. L. & Zei, G. (1967). Experiments with an artificial population. In *Proceedings of the Third International Congress in Human Genetics,* ed. J. F. Crow & J. V. Neel, pp. 473–8. Baltimore: Johns Hopkins Press.

Cavalli-Sforza, L. L. & Bodmer, W. F. (1971). *The Genetics of Human Populations.* San Francisco: W. H. Freeman.

Cherry, S. (1980). The hospitals and population growth: The voluntary general hospitals, mortality and local populations in the English provinces in the eighteenth and nineteenth centuries. *Population Studies,* 34, 59–75 and 251–65.

Cliff, A. D. & Haggett, P. (1984). Island epidemics. *Scientific American,* 250, 138–47.

Cliff, A. D., Haggett, P., Ord, J. K. & Versey, G. R. (1981). *Spatial Diffusion: An Historical Geography of Epidemics in an Island Community.* Cambridge: Cambridge University Press.

Cockburn, T. A. (ed.) (1967). *Infectious Diseases: Their Evolution and Eradication.* Springfield, IL: Charles C. Thomas.

Cockburn, T. A. (1971). Infectious disease in ancient populations. *Current Anthropology,* 12, 45–62.

Cohen, M. N. & Armelagos, G. J. (eds.) (1984). *Paleopathology at the Origins of Agriculture.* Orlando, FL: Academic Press.

Coleman, D. A. (1981). The effect of socio-economic class, regional origin, and other variables on marital mobility in Britain, 1920–1960. *Annals of Human Biology,* 8, 1–24.

Collins, J. J. (1982). The contribution of medical measures to the decline of mortality from respiratory tuberculosis: An age–period–cohort model. *Demography,* 19, 409–27.

Cook, S. F. & Borah, W. (1971). *Essays in Population History: Mexico and the Caribbean,* Vol. 1. Berkeley: University of California Press.

Cotter, W. (1967). Factors in the alteration of reproductive potential in chondystrophics. *Journal of Heredity,* 58, 59–63.

Crawford, M. H. (ed.) (1976). *The Tlaxcaltecans: Prehistory, Demography, Morphology, and Genetics.* University of Kansas Publications in Anthropology, Vol. 7. Lawrence, KS: University of Kansas Press.

Crawford, M. H. & Goldstein, E. (1975). Demography and evolution of an urban ethnic community: Polish Hill, Pittsburgh, *American Journal of Physical Anthropology,* 43, 133–40.

Crawfurd, M. D. (1972). A genetic study, including evidence for heterosis, of cystic fibrosis of the pancreas. *Heredity*, **29**, 126.

Crosby, A. W. (1972). *The Columbian Exchange: Biological and Cultural Consequences of 1492*. Westport, CT: Greenwood Press.

Crow, J. F. (1958). Some possibilities for measuring selection intensities in man. *Human Biology*, **30**, 1–13.

Crow, J. F. (1973). Some effects of relaxed selection. In *Proceedings of the Fourth International Congress on Human Genetics*, ed. J. De Grouchy, F. J. G. Ebling & I. W. Henderson, pp. 155–66. Amsterdam: Excerpta Medica.

Crow, J. F. (1980). The estimation of inbreeding from isonymy. *Human Biology*, **52**, 1–14.

Crow, J. F. (1983). Discussion (of 'Surnames as markers of inbreeding and migration'). *Human Biology*, **55**, 383–97.

Crow, J. F. (1989). Update to 'some possibilities for measuring selection intensities in man'. *Human Biology*, **61**, 776–80.

Crow, J. F. & Mange, A. P. (1965). Measurement of inbreeding from the frequency of marriages between persons of the same surname. *Eugenics Quarterly*, **12**, 199–203.

Darlu, P. & Cazes, M. H. (1988). Bias in the estimation of inbreeding coefficient and probability of origin of genes with errors in genealogy. *Human Biology*, **60**, 901–8.

Devor, E. J. (1979). Historical Demography in the Åland Islands, Finland: The Size and Composition of Households and Families in the Parishes of Finström and Kökar from 1760 to 1880. Ph.D. Dissertation, University of New Mexico.

Dinsmore, D. (1985). Demographic structure and opportunity for selection of Halfmoon Township, Pennsylvania: 1850–1900. *Human Biology*, **57**, 335–52.

Dobson, T. & Roberts, D. F. (1971). Historical population movement and gene flow in Northumberland parishes. *Journal of Biosocial Science*, **3**, 193–208.

Drake, M. (1974). *Historical Demography: Problems and Projects*. Walton Hall, Milton Keynes: The Open University Press.

Dyke, B. (1971). Potential mates in a small human population. *Social Biology*, **18**, 28–39.

Dyke, B. & Morrill, W. T. (1980). *Genealogical Demography*. New York: Academic Press.

Ellis, W. & Starmer, W. T. (1978). Inbreeding as measured by isonymy, pedigrees and population size in Torbel, Switzerland. *American Journal of Human Genetics*, **30**, 366–76.

Endler, J. A. (1986). *Natural Selection in the Wild*. Monographs in Population Biology 21. Princeton, NJ: Princeton University Press.

Eriksson, A. W. (1973). *Human Twinning in and Around the Åland Islands*. Helsinki, Finland: Commentationes Biologicae 64, Societas Scientiarum Fennica.

Epstein, E. & Guttman, R. (1984). Mate selection in man: Evidence, theory, and outcome. *Social Biology*, **31**, 243–78.

Firth, R. (1936). *We the Tikopia: A Sociological Study of Kinship in Primitive Polynesia*. New York: American Book Co.

Fisher, R. A. (1930). *The Genetical Theory of Natural Selection*. London: Clarendon Press.

Fix, A. G. (1975). Fission-fusion and lineal effect – aspects of the population structure of the Semai Senoi of Malaysia. *American Journal of Physical Anthropology*, **43**, 295–302.

Fix, A. G. (1978). The role of kin-structured migration in genetic microdifferentiation. *Annals of Human Genetics*, **41**, 329–39.

Fleury, M. & Henry, L. (1958). Pour connaitre la population de la France depuis Louis XIV. *Population*, **13**, 663–86.

Frankenburg, S. R. (1990). Kinship and mate choice in a historic Eastern Blue Ridge community, Madison County, Virginia. *Human Biology*, **62**, 817–35.

Fuster, V. (1986) Determinants of family size in rural Galicia (Spain). *International Journal of Anthropology*, **1**(2), 129–34.

Goodman, A. H., Thomas, R. B., Swedlund, A. C. & Armelagos, G. J. (1988). Biocultural perspectives on stress in prehistoric, historical, and contemporary population research. *Yearbook of Physical Anthropology*, **31**, 169–202.

Gottlieb, K. (ed.) (1983). Surnames as markers of inbreeding and migration. *Human Biology*, **55**, 209–408.

Gutman, R. (1956). The birth statistics of Massachusetts during the nineteenth century. *Population Studies*, **10**, 69–94.

Hagaman, R., Elias, W. S. & Netting, R. M. (1978). The genetic and demographic impact of in-migrants in a largely endogamous community. *Annals of Human Biology*, **6**, 505–15.

Halberstein, R. A. & Crawford, M. H. (1972). Human biology in Tlaxcala, Mexico: Demography. *American Journal of Physical Anthropology*, **36**, 199–212.

Hammel, E. A. & Howell, N. (1987). Research in population and culture: An evolutionary framework. *Current Anthropology*, **28**, 141–60.

Harding, R. M. (1985). Historical population structure of three coastal districts of Tasmania, Australia: 1838–1950. *Human Biology*, **57**, 727–44.

Harpending, H. (1974). Genetic structure of small populations. *Annual Review of Anthropology*, **3**, 229–43.

Harrison, G. A. & Boyce, A. J. (1972). Migration, exchange, and the genetic structure of populations. In *The Structure of Human Populations*, ed. G. A. Harrison & A. J. Boyce, pp. 128–45. Oxford: Clarendon Press.

Hed, H. M. E. (1984). Opportunity for selection during the 17th–19th centuries in the diocese of Linköping as estimated with Crow's Index in a population of clergymen's wives. *Human Heredity*, **34**, 378–87.

Hed, H. M. E. (1986a). Opportunity for Natural Selection in Sweden. Doctoral Thesis, Department of Genetics, University of Umeå, S-901 87 Umeå, Sweden.

Hed, H. M. E. (1986b). Selection opportunities in seven Swedish 19th century populations. *Human Biology*, **58**, 919–31.

Hed, H. M. E. (1987). Trends in opportunity for natural selection in the Swedish population during the period 1650–1980. *Human Biology*, **59**, 785–97.

Hed, H. M. E. & Rasmuson, M. (1981). Cohort study of opportunity for selection in two Swedish 19th century parishes with a survey of other estimates. *Human Heredity*, **31**, 78–83.

Henry, L. (1968). The verification of data in historical demography. *Population Studies*, 22, 61–81.

Henry, L. (1976). *Population: Analysis and Models*. New York: Academic Press.

Hinde, P. R. A. (1987). The population of a Wiltshire village in the nineteenth century: a reconstruction study of Berwick St James, 1841–71. *Annals of Human Biology*, 14, 475–85.

Hiorns, R. W., Harrison, G. A., Boyce, A. J. & Küchmann, C. F. (1969). A mathematical analysis of the effects of movement on the relatedness between populations. *Annals of Human Genetics*, 32, 237–50.

Hiorns, R. W., Harrison, G. A. & Gibson, J. B. (1977). Genetic variation in some Oxfordshire villages. *Annals of Human Biology*, 4, 197–210.

Hollingsworth, T. H. (1969). *Historical Demography*. London: The Camelot Press.

Howell, N. (1973). The feasibility of demographic studies in 'anthropological' populations. In *Methods and Theories of Anthropological Genetics*, ed. M. H. Crawford & P. L. Workman, pp. 249–62. Albuquerque: University of New Mexico Press.

Howell, N. (1986). Demographic anthropology. *Annual Reviews in Anthropology*, 15, 219–46.

Hudson, E. H. (1965). Treponematosis and man's social evolution. *American Anthropology*, 67, 885–901.

Hussels, I. (1969). Genetic structure of Saas, a Swiss isolate. *Human Biology*, 41, 469–79.

Jacquard, A. & Ward, R. H. (1976). The genetic consequences of changing reproductive behavior. *Journal of Human Evolution*, 5, 139–54.

James, A. V. & Morrill, T. (1990). Culturally irregular mating choices in a population isolate. In *Approache Pluridisciplinaire des Isolates Humains/ Pluridisciplinary Approach to Human Isolates*, ed. A. Chaventre & D. F. Roberts, pp. 333–8. Paris: Congres et Colloques No. 3.

Jorde, L. B. (1980). The genetic structure of subdivided human populations: A review. In *Current Developments in Anthropological Genetics*, Vol. 1, ed. J. H. Mielke & M. H. Crawford, pp. 135–208. New York: Plenum Press.

Jorde, L. B. (1984). A comparison of parent-offspring and marital migration data as measures of gene flow. In *Migration and Mobility*, ed. A. J. Boyce, pp. 83–96. London: Taylor & Francis.

Jorde, L. B. (1985). Human genetic distance studies: Present status and future prospects. *Annual Review of Anthropology*, 14, 343–73.

Jorde, L. B. (1989). Inbreeding in the Utah Mormons: An evaluation of estimates based on pedigrees, isonymy, and migration matrices. *Annals of Human Genetics*, 53, 339–55.

Jorde, L. B. & Durbize, P. (1986). Opportunity for natural selection in the Utah Mormons. *Human Biology*, 58, 97–114.

Jorde, L. B. & Lathrop, G. M. (1988). A test of the heterozygote-advantage hypothesis in cystic fibrosis carriers. *American Journal of Human Genetics*, 42, 808–15.

Jorde, L. B. & Morgan, K. (1987). Genetic structure of the Utah Mormons: Isonymy analysis. *American Journal of Physical Anthropology*, 72, 403–12.

Jorde, L. B. & Pitkänen, K. J. (1991). Inbreeding in Finland. *American Journal of Physical Anthropology*, 84, 127–39.

Jorde, L. B., Workman, P. L. & Eriksson, A. W. (1982). Genetic microevolution in the Åland Islands, Finland. In *Current Developments in Anthropological Genetics*, Vol. 2, ed. M. H. Crawford & J. H. Mielke, pp. 333–65. New York: Plenum Press.

Jorde, L. B., Pitkänen, K. J. & Mielke, J. H. (1989). Predicting smallpox epidemics: A statistical analysis of two Finnish populations. *American Journal of Human Biology*, 1, 621–9.

Jorde, L. B., Pitkänen, K., Mielke, J. H., Fellman, J. O. & Eriksson, A. W. (1990). Historical epidemiology of smallpox in Kitee, Finland. In *Disease in Populations in Transition*, ed. A. C. Swedlund & G. J. Armelagos, pp. 183–200. New York: Bergin & Garvey.

Kleinbaum, D. G., Kupper, L. L. & Morgenstern, H. (1982). *Epidemiologic Research: Principles and Quantitative Methods*. Belmont, CA: Lifetime Learning.

Kobayashi, K. (1969). Changing patterns of differential fertility in the population of Japan. In *Proceedings of the Eighth International Congress on Anthropological and Ethnological Science*, Vol. 1, pp. 345–57. Yeno Park, Tokyo: Science Council of Japan.

Koertvelyessy, T. (1983). Demography and evolution in an immigrant ethnic community: Hungarian Settlement, Louisiana, USA. *Journal of Biosocial Science*, 15, 223–36.

Kosten, M. & Mitchell, R. J. (1990). Examining population structure through the use of surname matrices: Methodology for visualizing nonrandom mating. *Human Biology*, 62, 319–35.

Kramer, P. L. (1979). Migration and the analysis of population structure: A study in the Åland Islands, Finland, 1750–1965. PhD Dissertation, University of New Mexico, Albuquerque.

Kramer, P. L. (1981). The non-Markovian nature of migration: A case study in the Åland Islands, Finland. *Annals of Human Biology*, 8, 243–53.

Küchemann, C. F., Boyce, A. J. & Harrison, G. A. (1967). A demographic and genetic study of a group of Oxfordshire villages. *Human Biology*, 41, 309–21.

Küchemann, C. F., Harrison, G. A., Hiorns, R. W. & Carrivick, P. J. (1974). Social class and marital distance in Oxford city. *Annals of Human Biology*, 1, 13–27.

Küchemann, C. F., Lasker, G. W. & Smith, D. I. (1979). Historical changes in the coefficient of relationship by isonymy among the populations of the Otmoor villages. *Human Biology*, 51, 63–77.

Lasker, G. W. (1954). Human evolution in contemporary communities. *Southwestern Journal of Anthropology*, 10, 353–65.

Lasker, G. W. (1985). *Surnames and Genetic Structure*. Cambridge: Cambridge University Press.

Lasker, G. W. (1988). Repeated surnames in those marrying into British one-surname 'lineages': An approach to the evaluation of population structure through analysis of the surnames in marriages. *Human Biology*, 60, 1–9.

Lasker, G. W., Mascie-Taylor, C. G. N. & Coleman, D. A. (1986). Repeating pairs of surnames in marriages in Reading (England) and their significance for population structure. *Human Biology*, 58, 421–5.

Lawton, R. (ed.) (1978). *The Census and Social Structure*. London: Frank Cass.

Lee, E. T. (1980). *Statistical Methods for Survival Analysis*. Belmont, CA: Lifetime Learning Publications.

Lee, R. (ed.) (1977). *Population Patterns in the Past*. New York: Academic Press.

Lees, F. C. & Relethford, J. H. (1982). Population structure and anthropometric variation in Ireland during the 1930's. In *Current Developments in Anthropological Genetics*, Vol. 2, ed. M. H. Crawford & J. H. Mielke, pp. 385–428. New York: Plenum Press.

Leslie, P. W. (1985). Potential mates analysis and the study of human population structure. *Yearbook of Physical Anthropology*, **28**, 53–78.

Leslie, P. W. & Gage, T. B. (1989). Demography and human population biology: Problems and progress. In *Human Population Biology: A Transdisciplinary Science*, ed. M. H. Little & J. D. Haas, pp. 15–44. Oxford: Oxford University Press.

May, R. M., Anderson, R. M. & McLean, A. R. (1988). Possible demographic consequences of HIV/AIDS epidemics. I. Assuming HIV infection always leads to AIDS. *Mathematical Bioscience*, **90**, 475–505.

May, R. M., Anderson, R. M. & McLean, A. R. (1989). Possible demographic consequences of HIV/AIDS epidemics. II. Assuming HIV infection does not necessarily lead to AIDS. In *Mathematical Approaches to Problems in Resource Management and Epidemiology*. Lecture notes in Biomathematics 81, pp. 220–47. Berlin: Springer-Verlag.

McCullough, J. M. (1989). Relation of community size to endogamy in traditional society: Pátzcuaro, Mexico, 1903–1932. *American Journal of Human Biology*, **1**, 281–7.

McCullough, J. M. & Barton, E. Y. (1991). Relatedness and kin-structured migration in a founding population: Plymouth Colony 1620–1633. *Human Biology*, **63**, 355–66.

McGrath, J. W. (1988). Social networks and disease spread in the Lower Illinois Valley: A simulation approach. *American Journal of Physical Anthropology*, **77**, 483–96.

McGrath, J. W. (1991). Biological impact of social disruption resulting from epidemic disease. *American Journal of Anthropology*, **84**, 407–19.

McKeown, T. (1976). *The Modern Rise of Population*. New York: Academic Press.

McKeown, T. (1979). *The Role of Medicine: Dream, Mirage, or Nemesis?* Princeton: Princeton University Press.

Meindl, R. S. (1984). Components of longevity: Developmental and genetic responses to differential childhood mortality. *Society Science and Medicine*, **16**, 165–74.

Meindl, R. S. (1987). Hypothesis: A selective advantage for cystic fibrosis heterozygotes. *American Journal of Physical Anthropology*, **74**, 39–45.

Meindl, R. S. & Swedlund, A. C. (1977). Secular trends in mortality in the Connecticut River Valley, 1750–1800. *Human Biology*, **49**, 389–414.

Menozzi, P., Piazza, A. & Cavalli-Sforza, L. (1978). Synthetic maps of human gene frequencies in Europeans. *Science*, **201**, 786–92.

Mercer, A. J. (1985). Smallpox and epidemiological-demographic changes in Europe: The role of vaccination. *Population Studies*, **39**, 287–307.

Mielke, J. H. (1980). Demographic aspects of population structure in Åland. In *Population Structure and Genetic Disorders*, edited by A. W. Eriksson, H.

Forsius, H. R. Nevanlinna, P. L. Workman & R. K. Norio, pp. 471–86. London: Academic Press.

Mielke, J. H. & Pitkänen, K. J. (1989). War demography: The impact of the 1808–09 war on the civilian population of Åland, Finland. *European Journal of Population*, **5**, 373–98.

Mielke, J. H., Workman, P. L., Fellman, J. O. & Eriksson, A. W. (1976). Population structure of the Åland Islands, Finland. *Advances in Human Genetics*, **6**, 241–321.

Mielke, J. H., Devor, E. J., Kramer, P. L., Workman, P. L. and Eriksson, A. W. (1982). Historical population structure of the Åland Islands, Finland. In *Current Developments in Anthropological Genetics*, Vol. 2, pp. 255–332. New York: Plenum Press.

Mielke, J. H., Jorde, L. B., Trapp, P. G., Anderton, D. L., Pitkänen, K. & Eriksson, A. W. (1984). Historical epidemiology of smallpox in Åland, Finland: 1751–1890. *Demography*, **21**, 271–95.

Mielke, J. H., Pitkänen, K. J., Jorde, L. B., Fellman, J. O. & Eriksson, A. W. (1987). Demographic patterns in the Åland Islands, Finland, 1750–1900. *Yearbook of Population Research in Finland*, **XXV**, 57–74.

Mørch, E. T. (1941). Chondrodystrophic dwarfs in Denmark. *Opera Ex Domo Biologiae. Hereditariae Humanae, Universitatis Hafniensis*, **3**, 1–200.

Morgan, K. (1973). Historical demography of a Navajo community. In *Methods and Theories of Anthropological Genetics*, ed. M. H. Crawford & P. L. Workman, pp. 263–314. Albuquerque: University of New Mexico Press.

Morton, N. E. (1969). Human population structure. *Annual Review of Genetics*, **3**, 53–74.

Morton, N. E. (1973). Prediction of kinship from a migration matrix. In *Genetic Structure of Populations*, Vol. III, ed. N. E. Morton, pp. 119–23. Honolulu: University of Hawaii Press.

Morton, N. E. & Hussels, I. (1970). Demography of inbreeding in Switzerland. *Human Biology*, **42**, 65–78.

O'Brien, E., Jorde, L. B., Rönnlöf, B., Fellman, J. O. & Eriksson, A. W. (1988a). Inbreeding and genetic disease in Sottunga, Finland. *American Journal of Physical Anthropology*, **75**, 477–86.

O'Brien, E., Jorde, L. B., Rönnlöf, B., Fellman, J. O. & Eriksson, A. W. (1988b). Founder effect and genetic disease in Sottunga, Finland. *American Journal of Physical Anthropology*, **77**, 335–46.

O'Brien, E., Jorde, L. B., Rönnlöf, B., Fellman, J. O. & Eriksson, A. W. (1989). Consanguinity avoidance and mate choice in Sottunga, Finland. *American Journal of Physical Anthropology*, **79**, 235–46.

Palmore, J. A. & Gardner, R. W. (1983). *Measuring Mortality, Fertility, and Natural Increase*. Honolulu: The East-West Center.

Petersen, W. (1975). A demographer's view of prehistoric demography. *Current Anthropology*, **16**, 227–47.

Pitkänen, K. J. (1977). The reliability of the registration of births and deaths in Finland in the eighteenth and nineteenth centuries: Some examples. *Scandinavian Economic History Review*, **25**, 138–59.

Pitkänen, K. J., Jorde, L. B., Mielke, J. H., Fellman, J. O. & Eriksson, A. W. (1988). Marital migration and genetic structure in Kitee, Finland. *Annals of Human Biology*, **15**, 23–34.

182   *J. H. Mielke and A. C. Swedlund*

Pitkänen, K. J., Mielke, J. H. & Jorde, L. B. (1989). Smallpox and its eradication in Finland: Implications for disease control. *Population Studies*, **43**, 95–111.
Pollard, A. H., Yusuf, F. & Pollard, G. N. (1990). *Demographic Techniques*. Australia: Pergamon Press.
Popham, R. E. (1953). The calculation of reproductive fitness and mutation rate of the gene for chondrodystrophy. *American Journal of Human Genetics*, **5**, 73–5.
Pressat, R. (1978). *Statistical Demography*. New York: St. Martin's Press.
Preston, S. H. (1976). *Mortality Patterns in National Populations with Special Reference to Recorded Causes of Death*. New York: Academic Press.
Ramenofsky, A. F. (1987). *Vectors of Death: The Archaeology of European Contact*. Albuquerque, NM: University of New Mexico Press.
Reddy, B. M. & Chopra, V. P. (1990). Occupation and the opportunity for natural selection: The Indian case. *International Journal of Anthropology*, **5**, 295–308.
Relethford, J. H. (1985). Examination of the relationship between inbreeding and population size. *Journal of Biosocial Science*, **17**, 97–106.
Relethford, J. H. (1986a). Density-dependent migration and human population structure in historical Massachusetts. *American Journal of Physical Anthropology*, **69**, 377–88.
Relethford, J. H. (1986b). Settlement formation in North-central Massachusetts, 1700–1850. *Social Biology*, **33**, 276–90.
Relethford, J. H. (1986c). The gravity model of human population structure. *Human Biology*, **58**, 801–15.
Relethford, J. H. (1988). Estimation of kinship and genetic distance from surnames. *Human Biology*, **60**, 475–92.
Relethford, J. H. (1991). Effect of population size on marital migration distance. *Human Biology*, **63**, 95–8.
Relethford, J. H. & Lees, F. C. (1983). Genetic implications of return migration. *Social Biology*, **30**, 158–61.
Roberts, D. F. (1968). Genetic fitness in a colonizing population. *Human Biology*, **40**, 494–507.
Roberts, D. F. (1971). The demography of Tristan da Cunha. *Population Studies*, **25**, 465–79.
Roberts, D. F. (1988). Migration and genetic change. *Human Biology*, **60**, 521–39.
Roberts, D. F. & Rawling, C. P. (1974). Secular trends in genetic structure: An isonymic analysis of Northumberland parish records, *Annals of Human Biology*, **1**, 393–410.
Roberts, D. F. & Sunderland, E. (eds.) (1973). *Genetic Variation in Britain*. London: Taylor & Francis.
Rogers, A. R. (1991). Doubts about isonymy. *Human Biology*, **63**, 663–8.
Rogers, A. R. & Jorde, L. B. (1987). The effect of non-random migration on genetic differences between populations. *Annals of Human Genetics*, **51**, 169–76.
Rogers, J. & Norman, H. (eds.) (1985). *The Nordic Family: Perspectives on Family Research*. Essays in Social and Demographic History, Reports from the Family History Group, Department of History, No. 4, Uppsala University, Sweden.

Rogers, L. A. (1987). Concordance in isonymy and pedigree measures of inbreeding: The effects of sample composition. *Human Biology*, **59**, 753–67.

Sattenspiel, L. (1987a). Epidemics in nonrandomly mixing populations: A simulation. *American Journal of Physical Anthropology*, **73**, 251–65.

Sattenspiel, L. (1987b). Population structure and the spread of disease. *Human Biology*, **59**, 411–38.

Sattenspiel, L. (1988). Spread and maintenance of a disease in a structured population. *American Journal of Physical Anthropology*, **77**, 497–504.

Sattenspiel, L. (1990). Modeling the spread of infectious disease in human populations. *Yearbook of Physical Anthropology*, **33**, 245–76.

Sattenspiel, L., Koopman, J., Simon, C. & Jacquez, J. A. (1990). The effects of population structure on the spread of the HIV infection. *American Journal of Physical Anthropology*, **82**, 421–9.

Shryock, H. S. & Siegel, J. S. (1980). *The Methods and Materials of Demography*, Vols. 1 and 2, Washington, DC: U.S. Department of Commerce, Bureau of the Census.

Smith, C. A. B. (1969). Local fluctuations in gene frequencies. *Annals of Human Genetics*, **32**, 251–60.

Smith, M. T. (1988). The Isle of Wight in 1851: Historical migration and genetic structure. In *Anthropologie et Histoire ou Anthropologie Historique?*, ed. Luc Buchet, pp. 113–22. Actes des Troisiémes Journées Anthropologiques de Valbonne, Notes et monographies techniques No. 24, Paris: Centre National de la Recherche Scientifique.

Smith, M. T. & Pain, A. J. (1989). Estimates of historical migration in County Durham. *Annals of Human Biology*, **16**, 543–7.

Smith, M. T., Williams, W. R., McHugh, J. J. & Bittles, A. H. (1990). Isonymic analysis of post-famine relationships in the Ards Peninsula, N.E. Ireland: Effects of geographical and politico-religious boundaries. *American Journal of Human Biology*, **2**, 245–54.

Spuhler, J. N. (1959). Physical anthropology and demography. In *The Study of Population*, ed. P. M. Hauser & O. D. Duncan, pp. 728–58. Chicago: The University of Chicago Press.

Spuhler, J. N. (1976). The maximum opportunity for natural selection in some human populations. In *Demographic Anthropology*, ed. E. Zubrow, pp. 423–51. Albuquerque NM: University of New Mexico Press.

Steinberg, A. G., Bleibtreu, H. K., Kurczynski, T. W., Martin, A. O. & Kurczynski, E. M. (1967). Genetic studies in an inbred human isolate. *Proceedings of the Third International Congress of Human Genetics*, ed. J. F. Crow & J. V. Neel, pp. 267–90. Baltimore: Johns Hopkins Press.

Sutter, J. & Tabah, L. (1955). L'evolution des isolates de deux departments Francais. *Population*, **22**, 709–34.

Svanborg-Eden, C. & Levin, B. R. (1990). Infectious disease and natural selection in human populations: A critical examination. In *Disease in Populations in Transition: Anthropological and Epidemiological Perspectives*, ed. A. C. Swedlund & G. J. Armelagos, pp. 31–46. New York: Bergin & Garvey.

Swedlund, A. C. (1972). Observations on the concept of neighborhood knowledge and the distribution of marriage distances. *Annals of Human Genetics*, **35**, 327–30.

Swedlund, A. C. (1975). Population growth and settlement pattern in Franklin and Hampshire counties, Massachusetts, 1650–1850. *American Antiquity*, **40**(2), 22–33.

Swedlund, A. C. (1980). Historical demography: Applications in anthropological genetics. In *Current Developments in Anthropological Genetics*, Vol. 1, ed. J. H. Mielke & M. H. Crawford, pp. 17–42. New York: Plenum Press.

Swedlund, A. C. (1984). Historical studies of mobility. In *Migration and Mobility: Biosocial Aspects of Human Movement*, ed. A. J. Boyce, pp. 1–18. London: Taylor & Francis.

Swedlund, A. C. & Armelagos, G. J. (1990). *Disease in Populations in Transition: Anthropological and Epidemiological Perspectives*. New York: Bergin & Garvey.

Swedlund, A. C. & Boyce, A. J. (1983). Mating structure in historical populations: Estimation by analysis of surnames. *Human Biology*, **55**, 251–62.

Swedlund, A. C., Temkin, H. & Meindl, R. (1976). Population studies in the Connecticut Valley: Prospectus. *Journal of Human Evolution*, **5**, 75–93.

Swedlund, A. C., Meindl, R. S. & Gradie, M. I. (1980). Family reconstitution in the Connecticut Valley: Progress on record linkage and the mortality survey. In *Genealogical Demography*, ed. B. Dyke & W. Morrill, pp. 139–55. New York: Academic Press.

Swedlund, A. C., Jorde, L. B. & Mielke, J. H. (1984). Population structure in the Connecticut Valley. I: Marital migration. *American Journal of Physical Anthropology*, **65**, 61–70.

Tempkin-Greener, H. & Swedlund, A. C. (1978). Fertility transition in the Connecticut Valley: 1740–1850. *Population Studies*, **32**, 27–42.

Trapp, P. G. (1987). Variability in natural fertility and the opportunity for natural selection in a preindustrial agricultural community: Finström, Åland Islands. Ph.D. Dissertation, Dept. of Anthropology, University of Kansas.

Upham, S. (1986). Smallpox and climate in the American Southwest. *American Anthropology*, **88**, 115–28.

van der Walle, E. (1976). The current state of historical demography. *Public Data Use*, **4**, 8–11.

Ward, R. H. & Weiss, K. M. (1976). The demographic evolution of human populations. In *The Demographic Evolution of Human Populations*, ed. R. H. Ward & K. M. Weiss, pp. 1–24. London: Academic Press.

Weiss, K. M. (1975). The application of demographic models to anthropological data. *Human Ecology*, **3**, 87–103.

Weiss, K. M. (1976). Demographic theory and anthropological inference. *Annual Review of Anthropology*, **5**, 351–81.

Weiss, K. M. (1989). A survey of human biodemography. *Journal of Quantitative Anthropology*, **1**, 79–151.

Weitman, S., Shapiro, G. & Markoff, J. (1976). Statistical recycling of documentary information: Estimating regional variations in a pre-censal population. *Social Forces*, **55**, 338–66.

Wijsman, E. M. & Cavalli-Sforza, L. L. (1984). Migration and genetic population structure with special reference to humans. *Annual Review of Ecology and Systematics*, **15**, 279–301.

Willigan, J. D. & Lynch, K. A. (1982). *Sources and Methods of Historical Demography*. New York: Academic Press.

Williams-Blangero, S. (1990). Population structure of the Jirels: Patterns of mate choice. *American Journal of Physical Anthropology*, **82**, 61–71.

Wood, J. W. (1986). Convergence of genetic distance in a migration matrix model. *American Journal of Physical Anthropology*, **71**, 209–19.

Wood, J., Smouse, P. & Long, J. (1985). Sex-specific dispersal patterns in two human populations of Highland New Guinea. *American Naturalist*, **125**, 747–68.

Workman, P. L. & Jorde, L. B. (1980). The genetic structure of the Åland Islands. In *Population Structure and Genetic Disease*, ed. A. W. Eriksson, H. Forsius, H. R. Nevannlinna, P. L. Workman and R. K. Novio, pp. 487–508. New York: Academic Press.

Workman, P. L., Mielke, J. H. & Nevanlinna, H. R. (1976). The genetic structure of Finland. *American Journal of Physical Anthropology*, **44**, 341–68.

Wrightson, K. & Levine, D. (1979). *Poverty and Piety in an English Village: Terling, 1525–1700*. New York: Academic Press.

Wrigley, E. A. (1972). *Nineteenth-Century Society: Essays in the Use of Quantitative Methods for the Study of Social Data*. Cambridge: Cambridge University Press.

Wrigley, E. A. (1977). Births and baptisms: The use of Anglican baptism registers as a source of information about the numbers of births in England before the beginning of civil registration. *Population Studies*, **31**, 281–312.

Wrigley, E. A. & Schofield, R. S. (1981). *The Population History of England, 1541–1871: A Reconstruction*. Cambridge MA: Harvard University Press.

Yasuda, N. (1975). The distribution of distance between birthplaces of mates. *Human Biology*, **47**, 81–100.

Yasuda, N. & Kimura, M. (1973). A study of human migration in the Mishima district. *Annals of Human Genetics*, **36**, 313–22.

Yasuda, N. & Morton, N. E. (1967). Studies on human population structure. *Proceedings of the 3rd International Congress of Human Genetics*, ed. J. F. Crow & J. V. Neel, pp. 249–65. Baltimore: Johns Hopkins.

8 *Writing for publication*

G. W. LASKER

Writing is part of research. It should not be put off until other aspects are done, but should commence early and continue throughout a research project. The working sequence normally starts with a research plan and a grant application, followed by interim reports if the grantor requires them. Then there is the research *paper*, the first full account of a new scientific finding. After that, new results will often be integrated with other information into symposia papers, book chapters and review articles. None of these tasks is peculiar to field and survey research in biological anthropology.

The advice one receives on writing may be diverse and can be confusing. A good rule to follow is always to accept the advice of the editor and follow the published instructions of the specific journal to which you submit an article. Advice from referees for the journal ordinarily take precedence over other sources, but the editor may not require that all points need be accepted. A knowledgeable colleague who has published effectively on similar subjects or special sources dealing with those subjects (such as this chapter) can be specific. Books concerning science writing, such as that of O'Connor (1991), while containing much worthwhile information, take account of practice in so many different fields of science and so many kinds of publication that they have to offer lists of alternatives about organization of the work, styles of referencing, and so on that may be at odds with practices in biological anthropology journals. Day (1989), however, has had long experience in editing biomedical journals and largely confines his attention to the writing of research papers *per se* and can be highly recommended in respect to that task. General works on English composition are of value only as a background for writing concise informative prose, and may contain recommendations that do not apply: for instance each time reference is made to a scientific concept, the term with that precise meaning should be used, but good English style may recommend against such repetitions. Perhaps more than other scientists, anthropologists attempt to emulate the style of non-scientific good writing, and this often has an adverse effect on the acceptability of their papers by the specialty journals.

Of the several thousand manuscripts that were submitted to *Human Biology* over 35 years, over half were marred by technical shortcomings in the presentation that could have been avoided by reference to the above mentioned sources of advice. It is very easy to go astray: inclusion of unnecessary material, failure adequately to specify the methods, and mixing up the discussion with the results or with the methods may lead, at least in a case that is viewed by the editor as borderline, to a rejection rather than a request for a revision. This chapter examines some of the problems met in writing for different types of publication and suggests ways of dealing with them. However, the 'Instructions to authors' and the style and form of presentation in the intended journal should always be followed in detail. Although editors are generally sympathetic to the writing problems of students, foreigners and others if they have a message of importance to convey, and the copy editor can be relied upon for help with such tasks as converting English to American spelling, efforts to get the most technical details of manuscript preparation right will always make a good impression and suggest that the same care may have been true of the research itself.

Oral reports

This chapter has to do with published writings, but it is important to realize that oral dissemination is also important and a formal oral presentation is taken to represent a completed written paper. The oral presentation at a meeting of a scientific society represents the first transmission of those findings and the abstracts of such meetings are the first publication of the results. In human biology the principal societies with sessions for contributed papers at annual meetings are the American Association of Physical Anthropologists in the U.S.A., the Society for the Study of Human Biology in the U.K. and corresponding national societies in Japan, Mexico, Spain and other countries. Membership is open to research students, who are encouraged to present their work, as well as to members of the profession. Papers presented in this way are considered to have received 'peer review' because: 1. The abstract will have been submitted in advance for review by experts and inclusion or exclusion from the program on the basis of scientific merit; and 2. such meetings are open for attendance by any member of the scientific community (usually on payment of a registration fee) and *the author is subject to questions and criticisms from the audience.* Of course the program organizers are faced with practical problems of engendering interest in their meetings and session chairpersons are faced with limitations of time and may cut off or preempt discussion; nevertheless, scientific communication hinges on the concept that communication should be open, criticism dispassionate, and the observed facts speak for themselves.

Some anthropologists make available full copies of their papers at this time, a practice borrowed from the social sciences. Such copies are sometimes marked 'not for publication'. The questions of whether such distribution compromises copyright and whether the disclaimer has any force are usually of no importance in respect to the kind of research biological anthropologists do, and one should feel free to distribute copies of draft manuscripts to colleagues for their comment. Some journal editors, especially of medical journals, look askance on any practice that smacks of prior publication, such as distribution of copies for use by the media, and there have been cases where unpublished results have been 'stolen', but usually one has the most to gain by the widest possible dissemination of one's results.

Other types of oral presentations, at international congresses, conferences, seminars, invited symposia, etc. are usually unsuitable for the primary presentation of research results because, unlike the society meetings, which alternate with each other and with those of local societies throughout the year, other types of meetings may be infrequent; also those attending may be more diverse, more specialized, or even restricted to an invited few.

Poster papers

Since there is not enough time available for all who wish to do so to give contributed papers from platform, the associations increasingly expect authors to give their papers in poster sessions. At the annual meetings of the Human Biology Council, for instance, the only contributed papers are in such sessions. The poster should consist of the title, 'by' and 'from' lines, the abstract, tables of data, figures (each with an explanatory legend), and a brief list of the chief findings. Most associations that hold poster sessions issue exact instructions concerning space available, times of setup and takedown, hours when authors should be present, etc. Some authors try to squeeze too much into their poster papers and make them hard to read. It is better to keep the type and other material at a large enough scale to be read at a distance of at least two meters. One is expected to be at the location of his or her poster to help explain it to those who come to see it. That gives an opportunity for questions, answers and amplification.

Research papers

It is the published paper rather than the orally presented research report that represents the hub of scientific communication, however. That is, the most important activity for initial standing as a research scientist is the report of original work in refereed journals devoted to the subject. Among the principal journals in English for reports of results of field work and survey studies in biological anthropology are the monthly *American Journal of Physical Anthropology,* and the bimonthlies, *Human Biology, Annals of Human Biology* and *American Journal of Human Biology.* There are many other journals that especially serve some part of the subject, such as the *Annals of Human Genetics* for population genetics and the *Journal of Biosocial Science* for social biology, and there may be advantages to publishing in the journal of the very subfield with which one wishes to be most identified. Trying to reach out to some audience one thinks *should* pay attention to the work often fails, but if the findings really have broad significance, a paper is a candidate for publication in one of the large circulation general science journals such as *Science, Nature* or the *American Scientist.* Also, the editors of such journals as the *American Anthropologist* cry out for the submission of manuscripts in the area of biological anthropology that would be of relevance to a broad audience of general anthropologists. Indeed, almost all editors think that articles of interest to a wider scientific audience would improve their journals. Alas, actual manuscripts about new original findings follow a different pattern since science itself generally proceeds by small increments – each new point being promptly disseminated as soon as it is observed. The increasing specialization of science is unlikely to be reversed by the creation of many manuscripts that are both original and wide ranging. It is almost always best to separate the two functions: to put the new material in brief research reports in specialized journals and to integrate these findings in articles in secondary publications directed at wider audiences.

As pointed out in Chapter 1 in respect to grant applications, a scientific presentation consists of fixed parts in a definite order: abstract, introduction, material, methods, results, discussion and literature cited. Efforts to improve on this system seldom succeed. There is usually a temptation to write as one thinks, following a train of thought, but editors of primary research journals almost always prefer one to stick to a single topic at a time and present it in the proper order.

However, there is no need to write a paper in that order. The title comes first, but is best written last. It should describe briefly what one did and found, but there is no need to try to squeeze in details of where the study was done and of what the sample was composed. Rather, the title should emphasize the main relationship, such as 'high versus low altitude', and if no difference was found, the word 'lack' (as in 'lack of relation of ABO and Rh+/− blood groups to body weight and height') will indicate the nature of the result. Except where previous positive assertions have been loudly acclaimed, however, negative findings have low priority for publication and such papers should be very short, under the dictum that for a paper to be acceptable, if it is only half the usual length it need seem to an editor to be only half as important. If the paper was given orally and later much revised, one should also reconsider the title.

The byline and fromline (or from-footnote) come next. They should follow the exact style of the journal to which the article will be submitted. All and only individuals who have contributed significantly to the planning, execution and analysis of the work are properly considered as authors. Furthermore, they should always be listed in the descending order of the significance of their contribution to the work. Note that those who merely contribute to the financial support are not thereby qualified to be authors. Purely technical assistance does not qualify one as an author of a scientific paper. It is getting the idea and demonstrating its occurrence that count. The fromline lists the institution that sponsored the work. If an author has changed employment he or she adds: 'Present address: ...' Readers appreciate a full mailing addresses in fromlines, but some journals publish less complete information about where the work is from.

The style may also permit a footnote or an endnote of the paper, entitled 'Acknowledgements'. This should be limited to financial sponsors plus those who have made other substantial contributions. Editors ask authors to keep the acknowledgments brief since this section is usually of interest mainly to the persons listed and is therefore considered to be an unimportant use of journal space. It is more direct to say: 'I thank ...' than 'I should like to thank ...'.

The abstract is a summary of the principal parts of the paper: material, procedures and results. Some abstracts merely list what papers are about, but that indicative type of abstract is less useful and to be avoided. The American National Standard for Writing Abstracts advised that the abstract should be informative and contain the principal information contained in the article or other document. Cremmins' (1982) book on abstracting is largely about how to read. If it is difficult to abstract a manuscript, the article itself may be poorly organized and in need for reworking to put the important information in a form and the places where it can easily be retrieved by readers. The abstract should avoid tables of data, figures and references to literature. Mathematical formulae are also best avoided. The reasons for these restrictions is that the abstract will be reprinted in *Biological Abstracts* and elsewhere without the 'References' section of the paper and also the author will have no opportunity to revise printer's proofs of the versions of the abstract reprinted in the abstract journals. Again, the journal that publishes the paper may have special requirements for length of abstracts. If the article has been revised after the publications of the abstract of the oral presentation or poster, the original abstract may no longer be entirely suitable and it should be revised.

A short list of *key words* usually accompanies the abstract and should include any index entries that will be of assistance to librarians and others in literature searches. There is no need to repeat words from the title among the key words since both the key words and a permutated index of the title will be used in sophisticated library searches. Thus the key words of an article about the height of Rabat one-year-olds might include: 'body-length', 'infant', 'Morocco' and 'stature'.

Materials and methods

One should always list the numbers of subjects. A statement that it was a 'random sample' is unsatisfactory unless the subjects were drawn from the total population of some category by a system using random numbers and the details are stated. Otherwise use of the word 'random' may merely suggest that the author paid little attention to possible biases and that inadequate attention was given to the method of recruitment of subjects. Usually subjects will have been recruited by techniques that it is hoped will minimize biases. If these techniques are spelled out it will allow the reader to form an independent judgment of the representativeness of the sample.

There is no need to state that 'subjects gave informed consent'. Of course ethical standards must be met, but readers of journal articles are the wrong people to be asked to police any such standards; it is the sponsors of the research who should try to do so. On the other hand, it is of scientific concern to know as much as is relevant about individuals who declined to participate so the reader can judge what this may imply about the extension to the general population of the findings on the subsample actually studied.

BOX 8.1.

Instructions to authors
A set of instructions appears in each issue of most journals and should be read before preparing the final draft of a manuscript. Part of that of the *Journal of Biosocial Science* is reproduced here by permission of the Biosocial Society.

Notice to contributors
Papers are considered for publication on the understanding that they have not been, and will not be, published elsewhere in whole or in part.

Manuscripts should be sent in duplicate to the Editor. They should be clearly typewritten, on one side of the paper only, with a 1 1/4 inch margin, and be double-spaced and in the English language. Spelling should follow that of the *Concise Oxford Dictionary*. The Editor reserves the right to make minor literary emendations in the final editing. The author is responsible for the accuracy of quotations, tabular matter and references. Manuscripts or tables prepared on dot-matrix printers are not acceptable.

Manuscripts should bear the title of the paper, name of the institution where the work was done, the present postal address of the author, if different from that of the institution, and a short running head of not more than 50 letters. Titles should be brief. A short summary should precede the text. Acknowledgements should be made in a separate section.

Diagrams must be numbered and should bear the author's name, short title of the paper and figure number on the back. Captions should be typed on a separate sheet.

Tables should be typed on separate sheets, be given Arabic numbers and be headed by adequate captions. Their approximate position in the text should be indicated by a note in the margin. They should not exceed in size the equivalent of one page of print. Weights and measures should be given in metric units. Standard abbreviations should be used (μg, mg, g, kg, ml, l, mm, cm, m, km, °C, %, <, >, hr, min). Abbreviations should not be followed by, and initials need not be separated by, full points (e.g., FPA). Mean values should, where possible, be accompanied by standard errors.

References in the text should be given in the manner that is standard in the Journal. Titles of journals should be abbreviated according to the *World List of Scientific Periodicals* and as also given in *World Medical Periodicals*. An unpublished paper should not be cited unless it is already in press.

Proofs will be sent to contributors for minor corrections, and should be returned to the Editor within one week. Major alterations to the text will be accepted only at the author's expense. The date of receipt given at the end of the paper is that on which the script as published was received or agreed.

If one begins to write down the *Introduction* and one's *Methods* while doing the work, a paper will be half written before the field work is over. A contemporary account of procedures will be more accurate than later memory will allow. However, it is likely to contain too much unnecessary detail and will probably need editing down to an appropriate length. Annotated field protocols may provide examples to illustrate how the work was actually accomplished. If alternative methods were tried and some found wanting or abandoned, it rarely needs be mentioned in the methods section; if the points are important enough to make at all, they are more appropriate for the discussion section.

The *Results* section is best written by direct reference to tables of data and graphs of the tabulated data. Most experienced authors lay all their tables of data and analyses out in front of them as they write this section. Then they include in the Results section of the paper only the tables and charts that they need and actually cite. Editors do not like repetition of the exact same information in tables, graphs and text nor in any two of these. It may be better to show all of a large number of data points on a graph, the means, standard deviations and other statistical constants in a table, and the salient features and interpretation of the statistics in the text, this may give a much fuller understanding of the results than if one crams all of these into each type of presentation.

Journal editors have some notions about the numbers of figures, amounts of tabular material and total length appropriate for an article. These may not be stated but can be inferred from past issues of their journals. One should try to limit one's own manuscripts in a similar way because, as mentioned before, editor's standards of acceptance vary by article length. The best way to conserve space is not only to avoid direct duplication, but also to avoid giving information that can be directly inferred from other data; thus if numbers and standard deviations are given, there is no need for standard errors. Another great saving is to avoid discussion of findings that are not statistically significant; what one might say on such subjects is less likely to hold up on further study and even if it should prove to be right, one deserves little credit for saying so before there was enough evidence to show it.

Even the *Discussion* should stick to the main point. Here is the place, however, to relate one's work to what has previously been published. Comments about future work also may be included in the discussion. A considerable number of published articles would have been even better if some of the material in their introductions have been placed in the discussion. The same is true, perhaps to a lesser extent, of some comments in the results and methods sections. Some authors also wish to report *Conclusions*. Properly, a scientific study does not 'conclude', since all findings must be held with a certain reserve and subject to revision to the extent that future work fails to confirm. A short listing of findings may be recapitulated at the end of the discussion, but this must be brief since the interests of conciseness dictate a minimum of repetition.

The *References* list only (and all) works cited in the text and should be in the style of the journal, double spaced and carefully checked. A carefully inserted correction is far preferable to a published error. Only published and accepted 'in press' materials are suitable for inclusion as references. Other types of documentation, such as 'personal communication', are cited within the body of the paper with enough information about the sources for an interested reader to be able to locate them.

Refereeing

It is best to keep a copy of each manuscript in the *exact* form in which it was submitted. If there have been even minor corrections and one retains only the prior draft with inserted corrections, it may later be difficult to locate the places to which referees and the editor refer in their criticisms. Furthermore, when the editor receives copies of a manuscript, the intellectual content is the author's, but the physical copies belong to the journal and they may not be returned.

Editors of primary research journals file one copy (with the original illustrations) and submit copies to at least two referees. Therefore, one should send three copies unless a different number is specified in the instructions to authors. Scientists owe the duty of occasional refereeing as a function of their membership in the profession. Thus their responsibility is more to the journal and the potential readers than to the author. Nevertheless, they should, and usually do, try to be helpful to the author. The review usually consists of two parts: 1. Detailed criticisms on a marked copy, or separately where they are indexed by citation to the page and line to which they refer; and 2. A recommendation concerning the acceptability of the manuscript as a whole. The second part may involve considerations about the journal as well as about the manuscript and, in any case, is often not made available to the author, so what follows concerns only the criticisms of the content and style. Comments about these are usually transmitted anonymously to the author, with a statement of whether the article is accepted, accepted subject to certain conditions, or rejected. It is well to consider each and every point in the criticisms. The referees have been selected because the editor believed that they were better informed on the subject than the average reader of the journal. Also, the reviewers usually take more care in reading the manuscript than will most later readers of the published article. That is, if a referee or the editor misunderstands a point, an average reader is at least equally likely to misunderstand it. Therefore, virtually every criticism that bears on presentation requires some response by a change in the manuscript (although not necessarily the exact change that a referee may recommend).

Criticism of the work itself is more difficult for an author to deal with. If further analysis is called for, it is usually possible to comply. If there is a call for more information the author may have it available. If the need is for different procedures or more data, it may be too late to remedy, especially in the case of field work. However, if one pursues further the effort to publish (whether in the same journal or another) a study in which the critics have described the data themselves as inadequate, one should always try to do something about the criticism. Even if the flaws can not be eradicated, a frank admission of them may leave enough of interest to warrant publication. Again, brevity is especially important if there appears to be some such shortcoming.

At any stage one can consult critics of one's own choosing. Most biological anthropologists willingly give time to the consideration of the efforts of fellow research workers. Few are bothered by requests for help if they feel that the authors have already tried to do their best and have reread and revised their manuscripts before asking for aid. A reading by, and criticisms from, a colleague one knows permits discussion and may be more satisfactory than relying entirely on written criticisms from anonymous referees.

Most editors want nothing more to do with a rejected manuscript, but an author who communicates about a rejected manuscript can sometimes be given additional encouragement about how to proceed with a different paper based on the same study or about some other journal for which the article or a revision of it might be suitable. That is, there are some items of advice that an editor may be reluctant to volunteer, but that can be elicited by a request during a give and take discussion such as a telephone conversation.

If an article is accepted subject to certain conditions (at least minor conditions are attached to the majority of acceptances), it is almost always better to try to meet the conditions than to start over with a different journal. Sometimes authors misread as a 'condition' what the editor presented merely as a 'suggestion'. Under those circumstances the withdrawal of the article may be resented since the journal has already invested efforts in its review. The appearance of such an article in a second journal when there has been no formal withdrawal from the first may be misinterpreted as a 'multiple submission', an act that is strictly taboo.

Other types of publication

There are many other types of scientific communications besides the research paper and each type has a different function. All these types are grouped under the rubric 'secondary publications'. It is best not to put the first report of new research results in a periodical that is devoted to secondary publication or in an article that looks like a secondary publication, because the likelihood that it will be found and recognized in a literature search is less than would be true for research papers in research journals. Secondary publications include reviews of published literature, general articles such as those published in journals of sister sciences to stimulate cross disciplinary interest, chapters in books relating the topic of research to the theme of the book and contributions to symposia and organized sessions of international congresses (such as the International Congress of Anthropological and Ethnological Science) or other meetings. Some are more like the non-fiction products of science writers. Thus secondary publications are very diverse in purpose and style and there is probably little useful guidance that can be applied generally.

Review articles often help one find and distil the prior literature. Furthermore, writing a review article is a good way to familiarize oneself with the literature and, when it is completed, that task offers closure for one phase of preparing for a research project. There are several specialized journals for reviews (e.g. the *Yearbook of Physical Anthropology, Quarterly Review of Biology,* and *Annual Reviews of Anthropology*), and many of the research journals occasionally publish review articles. A review article by an applicant for a research grant is evidence that part of the preparatory work has already been completed and that the investigator is capable of carrying through projects to publication. It also is a way to identify the remaining problems and hence point the way for new research.

One should not denigrate the importance of publications such as reviews that may appear in non-refereed journals and that include no new findings. For individuals who have already established their place in human biology through publication of papers in the specialized literature, further enhancement of their reputations comes precisely from their books and other more general writings. Good biological anthropology research is of interest to a wider audience.

One's work may have an importance for public policy that will be missed or delayed unless the results are disseminated through the media. Reporters are likely to approach research work from a very different angle than scientists, and may sometimes misunderstand or sensationalize some aspect. Most North American universities have a department with staff members specialized in public relations; the staff have been hired because of their understanding of media requirements, but, unlike reporters, the public relations staff have a primary responsibility to promote the good name of their institution and hence they will wish the press reports reliably to reflect the work of the university. Thus publicity for one's research is desirable, but a wise first step in seeking it is an interview with a public relations professional. Such a PR officer will give one an opportunity to check and correct the press release. However, of course, he or she cannot determine whether the story will be used or what aspect of it will be played up. Reporters usually consult and quote several authorities with diverse opinions and do not use press releases in their entirety.

The use of computers

Computers are useful in virtually every activity of doing research. Although potentially one can integrate all stages of a project through computer use, this is usually not practicable and is not necessary. The system for the exchange of scientific knowledge is necessarily geared to accommodate those who use computers only in limited ways and therefore all the journals publishing original papers in biological anthropology still require submission of conventional hard copy manuscripts. Thus the authors' use of computers is disaggregated from the publishers'.

All of the following capacities may apply to research, and some of them are usually used in relation to publication.

1. Library searches with *Medline* or other indexes.
2. Citation management systems such as *Procite* or *Endnote* for handling and editing references.
3. Data recording with online recording instruments and analog to digital transcription.
4. Data management with *Dbase* or spreadsheet programs.
5. Communications via *E-mail* or other networks.
6. Statistical analysis using such programs as *SPSS* or *SAS* to perform univariate and multivariate analyses.
7. Preparation of graphs with graphing programs or facilities bundled into a word processing or a statistics program.
8. Word processing to produce manuscripts.

9.   Production of other illustrations and complex formulae in camera-ready form with programs for these purposes.
10. Use of editing facilities, spell-checking, dictionaries, thesauruses and grammar and style aids.

Presentation graphics software are not mentioned above because they pertain primarily to platform talks and poster presentations. Desktop publishing programs such as *PageMaker* with their numerous fonts and typesetting capacities are also not usually applicable to the author's role in publication.

Programs are constantly being updated and replaced, so the following comments about specific programs are merely examples of some software currently in use. Most canned programs applicable to writing for publication perform more than one task and a good tool for one purpose may be associated with a less satisfactory tool for another. Since separate programs for each tool involve a waste of effort, however, convenience (and cost) dictate a compromise. It is still almost always best to use printed books for facilities that are only rarely used, such as foreign language dictionaries, technical glossaries, atlases and style manuals.

The most important computer tool for manuscript preparation is the word processor. However, one often needs to alter or override the defaults because, for manuscripts to be set in type by the publisher, editors prefer:

1.   margins to be wide.
2.   Full double spacing (3, not 4 or 5 lines per inch) including the References but not the tables.
3.   Automatic hyphenation suppressed with no lines ending with a hyphen.
4.   Words to be set in italics underlined.
5.   Paragraphs always indented.
6.   Pages numbered.
7.   Copies made on a copier, not by computer so that page and line numbers correspond in all copies, including those retained by authors, and so that any hand-lettered symbols or insertions are identical in all copies. (An error corrected by Lasker and Raspe, 1992, was caused by the first of these authors having overlooked the absence of such a hand-lettered insertion in a computer-generated copy).

By preparing their own manuscripts with a word processor, researchers ought to be able to free support personnel from the task of typing for the more valuable service of editing. All too often the editing has been too perfunctory and, despite the availability of spell checks and other computer tools, the number of spelling, typographical, grammatical and other errors appearing in such journals as the *American Journal of Physical Anthropology* has not noticeably decreased as the use of computers for word processing has increased.

The computer may introduce new types of errors. For instance, 'The' will become 'he' if the control key rather than the shift key is depressed with the 't', and since 'he' is a word, the spell-checker will not highlight it. Also a symbol on the keyboard may not be the same as that understood by the printer.

Rabinovitz (1991) recently reviewed computer 'writer's tools'. The facilities tested and evaluated include grammar checker, dictionary, thesaurus and spell-checker. Rabinovitz first lists which of these facilities is bundled into each of the word-processor programs. He then rates on a four-point scale the presently available tools for the PC. For instance, *PowerEdit* is his best-rated stand-alone grammar-checker, but the test showed that it flagged more non-errors than some other programs. The bundled packages that do all four listed tasks are not completely integrated, but Rabinovitz considers *The Writer's Pack* to be a good bargain despite its lack of a thesaurus. The list of words used to test the performances of the dictionaries, spell-checkers and thesauruses reveals common types of errors, but could be modified or expanded to meet the needs of someone testing programs for their type of special vocabulary.

Next to word-processing and writer's tools, graphing is the most important computer facility for the preparation of manuscripts by human biologists. In conjunction with a laser printer, programs such as *Sigma Plot, Coral Draw, 35 Express, Harvard Graphics, Cricket Graph* (for the Mac) and *Telegraf* (for the main frame) are all said to produce figures ready for the lithographer's camera, but if the journal sets legends in type, it may be necessary to modify the printout with white ink to meet the requirements of the journal. Many kinds of illustrations are still best left to professional illustrators, although with programs for introducing standard elements into drawings, and others with base maps, some illustrating can be done by the investigators themselves.

Presentation graphics programs are chiefly of use for oral presentations. Fridlund *et al.* (1991) tested seven software programs for performance, presentation organization, text charting, numeric charting, editing capabilities, presentation tools, output, documentation, ease of learning, ease of use, support policies, technical support and value. *Aldus Persuasion 2.0* received the highest overall score among programs run on

Windows and *DrawPerfect 1.1* the highest score for those run on DOS, but the rating table is so arranged that one can easily introduce different weightings of the 12 criteria with possibly different conclusions.

Publishers, especially of monographs, may place some of the publishing tasks back on the researcher. If camera-ready copy is required, such programs as *PageMaker* will help. If copy on a diskette is required, it may be best to convert to an ASCII file oneself so the text can be edited before transmission. Lastly, in desperate situations where all that one has is a hard copy, it is possible to read it back into a computer file by use of a scanner and appropriate software. Like all other output, the eventual manuscript needs to be reread and edited.

References

Cremmins, E. T. (1982). *The Art of Abstracting*. Philadelphia: ISI Press.

Day, R. A. (1989). *How to Write and Publish a Scientific Paper*. 3rd edn. Cambridge: Cambridge University Press.

Fridlund, A. J., Zittle, T., Van Cura, J. & Kaliczak, A. (1991). Presentation graphics for DOS and Windows. *Infoworld*, **September 16**, 61–75.

Lasker, G. W. & Raspe, P. D. (1992). Given name relationships support surname 'genetics'; a note and correction. *Journal of Biosocial Science*, **24**, 131–3.

O'Connor, M. (1991). *Writing Successfully in Science*. London: Harper Collins Academic.

Rabinovitz, R. (1991). Writer's tools. *P. C. Magazine*, **10(16)**, 321–69.

Index